Clarence Edgar Edwords

Camp Fires of a Naturalist

The Story of Fourteen Expeditions after North American Mammals ...

Clarence Edgar Edwords

Camp Fires of a Naturalist
The Story of Fourteen Expeditions after North American Mammals ...

ISBN/EAN: 9783337250836

Printed in Europe, USA, Canada, Australia, Japan

Cover: Foto ©berggeist007 / pixelio.de

More available books at **www.hansebooks.com**

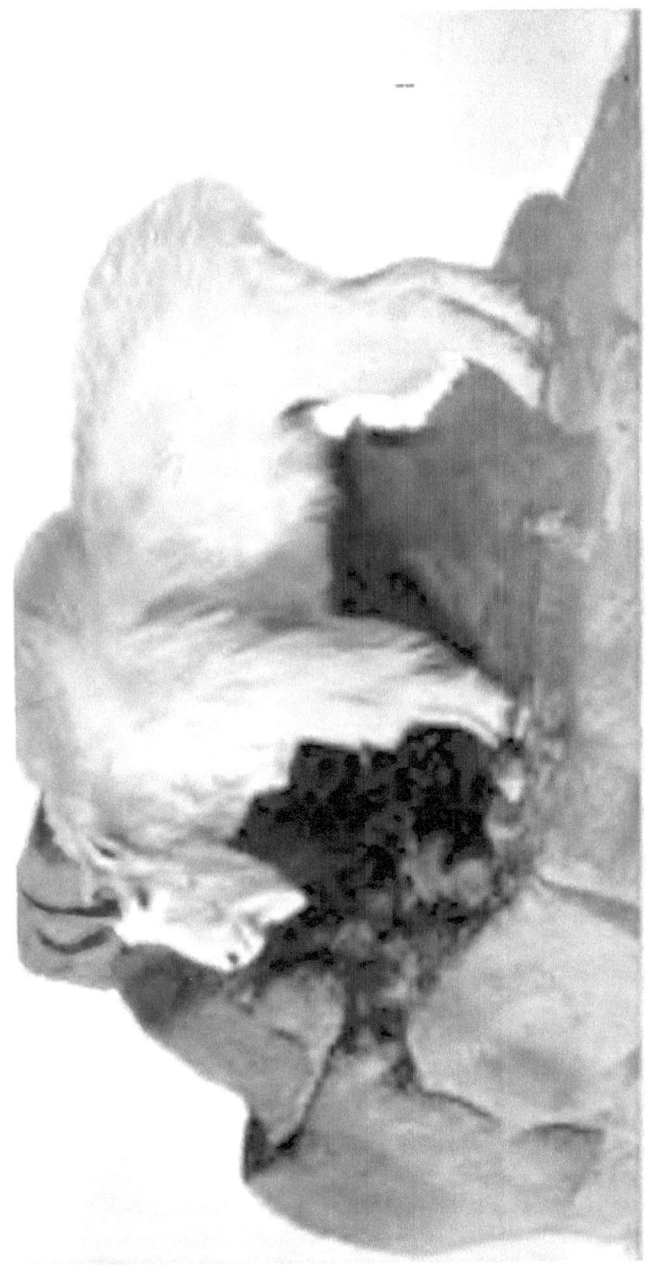

The Rocky Mountain goat.

CAMP-FIRES
OF A NATURALIST

THE STORY OF FOURTEEN EXPEDITIONS AFTER NORTH
AMERICAN MAMMALS, FROM THE FIELD NOTES OF
LEWIS LINDSAY DYCHE, A.M., M.S., PROFESSOR
OF ZOÖLOGY AND CURATOR OF BIRDS
AND MAMMALS IN THE KANSAS
STATE UNIVERSITY

BY
CLARENCE E. EDWORDS

ILLUSTRATED

NEW YORK
D. APPLETON AND COMPANY
1893

WITH A DEEP SENSE OF GRATITUDE

THIS BOOK IS DEDICATED

TO A

TRUE FRIEND, A CAREFUL CRITIC, AND A WISE COUNSELLOR,

CHARLES S. GLEED,

WHO, BY WORD AND DEED, HAS HELPED THE WRITER OVER THE

ROUGH PLACES AND MADE SMOOTH HIS PATHWAY

IN PLACING THIS BOOK BEFORE THE PUBLIC.

PREFACE.

HUNTING-STORIES without exciting adventures and hairbreadth escapes find little favour with the general public, but those who have spent pleasant hours in the woods know that such adventures are of rare occurrence.

This book deals solely with facts. The adventures of Professor Dyche while collecting specimens of the larger mammals of North America are not of a thrilling kind, but they give the life in the woods as it really is. My story is taken from the note-books and diaries of the professor, and not one word has been added to the facts which he has noted, to make the book more interesting or exciting.

I offer simply a description of the life of a naturalist-hunter and of the pleasures of camp life, with a naturalist's explanations of the habits of animals.

CLARENCE E. EDWORDS.

CONTENTS.

CHAPTER I.

Nursed by a Squaw—Boyhood and Early Manhood—How an Education was Obtained, ... 1

CHAPTER II.

Old Jim Leatherman's Load—The Storm—Some Experiences with Snakes—About Antelope—Useless Tails and Queer Horns—Fight with a Rattlesnake—How Dyche Scared the Tramps out of Camp, ... 5

CHAPTER III.

In New Mexico—Encounter with a Bear—Cold Hospitality—The Retired Boston Merchant and his Mountain Ranch—An Able Missouri Liar—The Mule that was not a Mule—Seven Deer with Seven Shots, . . 21

CHAPTER IV.

In New Mexico—Hunting Bear—Once More at Harvey's—The Same Luck—Camp Bear Trail—The Last Pot-Shot—A Long Chase—In Ambush—Too Much of a Good Thing—A Monarch Dethroned—What was Done with the Royal Hide, ... 39

CHAPTER V.

Making a New Trail—The Storm on the Mountain-Top—Neither Bear nor Elk—A Journey in Utter Darkness—Catching Foxes—Unwelcome Visitors—Brown's Avalanche—The Bear was Handicapped—An Experience with Ants and Fleas, ... 57

CHAPTER VI.

A Peculiar Wedding-Journey—Field Taxidermy—A Typical Mountaineer—Big Bear Talk—The Cabin was Looted—A Lost Timepiece—A Great Day for Talking, 76

CHAPTER VII.

In Indian Territory—A Frontier Fiddler—Life in a Dugout—Wild Turkeys, Wildcats, and other Wild Things—Surviving a Blizzard—An Indian Dance—An Embarrassed Professor—Successful Hunting—The Extinction of the Wild Turkey, ... 97

CHAPTER VIII.

In the Cascades—Extermination of Many Species of Animals—Something about the Rocky Mountain Goat—An Arduous Journey—The Cascades Reached—Wholesale Hunters—In Camp—A Failure, . . 118

CHAPTER IX.

The First Big-Horn—How his Skin and Bones were Prepared—Habits of the Rocky Mountain Sheep, . . 134

CHAPTER X.

End of Sheep-Hunting—How the Sentinel Fell at his Post—A Peculiar Wound—Finding the Noon Hour by Stars—How the Collection of Sheep was Completed, . 147

CHAPTER XI.

On Kettle River—Okonagan Smith and his Lonely Ranch—The Great Northern Boundary—Trials and Tribulations—"Fool Hens"—Through Fallen Trees—An Arduous Journey, ... 162

CHAPTER XII.

In the Cascades—A Forest Fire—After Rocky Mountain Goats—The First Shot—Down the Mountain—A Successful Hunter—A Night of Hardships—A Naturalist's Labours, ... 174

CONTENTS.

CHAPTER XIII.

A Peculiar Danger—Four Goats in Four Shots—A Rapidly Disappearing Tribe—Description and Habits—A Persistent Hunter, ... 190

CHAPTER XIV.

On Kettle River—Two Model Camp Companions—A Royal Deer—Eating Beaver Tails—A Tramp over the Mountains—Wolves about the Cabin—Varieties of Deer, . 204

CHAPTER XV.

At the Lake of the Woods—After Moose—A Plague of Mosquitoes—Dark Swamps and Deep Rivers—Compensations, ... 224

CHAPTER XVI.

In the Swamps—Habits of the Moose—The Moose-Call—On the River—Good Shooting Secures a Group—The King of Game Animals—The Naturalist nearly Killed, . 236

CHAPTER XVII.

With the Indians—How Indians Hunt Big Game—The Parallel Trails—Indian Superstitions—A Potent Beverage—Moose all "Nickoshin"—Return to Civilisation, ... 257

CHAPTER XVIII.

In Colorado—On the Trail of Elk—A Night in the Snow—Deer, but no Elk—Another Wild-Goose Chase—The World's Fair King—The Last Hunt, ... 276

CHAPTER XIX.

Results of the Camp-Fires—The Specimens Obtained, . 300

LIST OF ILLUSTRATIONS.

	FACING PAGE
The Rocky Mountain goat	*Frontispiece*
A corner of the storeroom	1
A family group	5
A mule deer buck	21
As he appeared in the opening	39
"They are sure good bear knives"	57
A group of Virginia deer	97
On guard	118
Almost despairing	190
Off for a long tramp	204
Ready to go home	213
The monarch of the woods	224
A mountain king	276

A corner of the storeroom.

CAMP-FIRES OF A NATURALIST.

CHAPTER I.

Nursed by a Squaw—Boyhood and Early Manhood—How an Education was Obtained.

ONE raw March evening, in the early days of Kansas, a covered wagon drawn by oxen stopped on the bank of the Waukarussa River. In the wagon lay a babe close to the side of its mother, whose illness was so severe that but little attention could be paid to the child. The sturdy pioneer, who had left his Eastern home to make a new one in the West, cheered his wife with a word and placed the child on a bed of grass before a bright fire. Near the spot selected for the camp was the winter village of a band of Indians, and the fire had been hardly started when a number of the red men gathered around the wagon. The condition of the sick mother appealed to the womanly instincts of the squaws, and tender hands ministered to her wants. The infant was taken from its improvised bed, and soon was drawing a new life from a red breast. For weeks the mother hovered between life and death, and all the while the babe was cared for in the village of the Indians. He thrived, and when the mother was restored to health the baby boy was strong and lusty.

This babe was Lewis Lindsay Dyche and his life almost began at a camp-fire. With the precocity often seen in pioneer life he seemed to pass from infancy to manhood with no intervening period of boyishness or youth. All the pleasures of his tender years were combined with business. At the age of nine he was hunting and trapping along the banks of the Waukarussa. His playmates were his dogs; his playthings were the beasts and birds; his playgrounds were the woods and prairies and the camps of the Indians. His hard lot and that of his father and mother taught him the value of money. Work was as natural to him as play to ordinary boys. For a five-cent piece he would follow the horses of the sorghum mill all day long, and this money would be hoarded with that received for the furs obtained during the winter's trapping and hunting.

All this time the alphabet was a mystery to him, and while he was in demand among the neighbors as a worker, he realized that to succeed in life, even among frontiersmen, it was necessary for him to have other learning than that obtained in the woods. At the age of sixteen he was tall and well formed, with the habits and appearance of a man. He had learned the rudiments of reading at the age of twelve, but shame prevented the tall, rawboned boy from showing his ignorance in the village school, and he advanced slowly in his learning. At the age of sixteen he found that his hoarded money had accumulated until he was the possessor of $600.

With **this** he determined to obtain an education, and for three years he studied at the State Normal School at Emporia, Kansas, preparing his mind to receive that higher education which was gradually opened to **him.** In order to husband his means, he rented a small room and did his own cooking during the three years. Mrs. A. P. Morse, an instructress in the school, took a deep interest in the young man, and **gave** him many valuable suggestions regarding his mode of study, which helped him over the rough path**ways** until he was able to travel them alone.

In addition **to** the mild persuasiveness of Mrs. Morse, there came a help of a different nature. Dr. C. R. Pomeroy, that sturdy and resolute president of the State Normal School whose strength of character has impressed so many minds in the West, saw the ambition of the diffident young backwoodsman, and opened to his mental vision such vast fields of research that he breathed a new life into the veins of the young man.

After the three years at the normal school, he, with one companion, went in a covered wagon to Lawrence, where the Kansas State University is situated, and there they camped in a sheltered vale just north of the university buildings. At night they slept in their wagon, and their meals were cooked at a camp-fire. They camped here until the cold weather drove them to warmer shelter, and then they rented a small room in the city, and there did their own cooking for the remainder of the year.

While the work at the normal school had been hard because it was strange, here it was **hard** because it was made so by the enthusiast. Night and day he pored over his books, only leaving them when dragged away by his fellow-students, who saw that he was breaking himself down by his close application. But even his recreation was work. He wandered over the fields and along the river, collecting insects for Professor Snow, and became so expert that he was afterward regularly employed by the professor to assist him in making his collection. From this source he obtained a considerable sum of money, which went toward defraying his expenses.

The university was then in its infancy, and Dyche saw that by properly combining his studies he could master more than one course at a time, and to this end he devoted himself. So well did he apply his vigorous mind that at the end of five years he graduated at the head of his class, obtaining two degrees. Nor did he stop here, for when he was preparing for new fields, Professor Snow offered him the position of assistant in his department. Under the guidance of the professor the young man became invaluable to the university, and the chair of anatomy and physiology was created and given to Dyche. Later on the branches of zoölogy, animal histology, the curatorship of birds and mammals, and the practical work of taxidermy were placed under his direction and guidance, with a corps of assistants.

The museum of the university, where stands the finest collection of mounted animals in the world, tells how well he has done his duty, and to the story of the incidents and adventures of the young man's life while in search of these mammals the pages of this book are devoted.

A family group.

CHAPTER II.

Old Jim Leatherman's Load—The Storm—Some Experiences with Snakes—About Antelope—Useless Tails and Queer Horns—Fight with a Rattlesnake—How Dyche Scared the Tramps out of Camp.

T was a jolly quartette that rode in old Jim Leatherman's wagon over the prairies of western Kansas, **on a hot** July afternoon, **from** Buffalo, a little shipping station **of the Union Pacific Railroad.** A casual observer would have **taken the party** for a lot of schoolboys out on a frolic, and he would not have been **far** wrong. The two older boys were the jolliest of all. The younger boys **were not** quite so demonstrative, yet they, too, were full of animal life and **were** inspired by the invigorating air of the plains.

The old boys called each other **Mudge and Snow**, while the younger ones **were** known as Dyche and Dick. **At** home Mudge and Snow were staid college professors, and Dyche and Dick were students in the institution. While **in** college circles decorum was duly observed, here on the prairies all four called each other by the most convenient names, and while they are removed from **the** college atmosphere these names may be sufficient.

Old Jim Leatherman found his strange load incomprehensible. Mudge **would** suddenly break off from his rhapsody on the possible skeleton of a plesiosaurus that he expected to find, jump from the wagon, and scurry over the prairie after Snow. Then both would return, triumphant over the capture of some inoffensive insect which the bright eyes of the old boys had detected as they rode along.

The day thus passed was short to the travellers. Many trophies of the "bug hunters" had been gathered when the wagon halted by the side of a little stream which wound across the barren plain. A lone cottonwood tree stood sentinel on the bank as a beacon to the party, and Leatherman interrupted a scientific discussion with the announcement:

"Here's the place and there's your firewood," pointing to the tree.

The tent was soon pitched, a blazing fire started, a pan of bacon set to frying, and the camp life of the naturalists was begun in earnest. Each had an assigned task. One got wood, one cooked, and the others arranged the tent and camp-ground. The sunset presaged foul weather. The whole horizon was banked with clouds. Snow and Dyche took up their quarters in the tent, while Mudge, Dick, and Leatherman spread their blankets on the sand at the foot of the sentinel cottonwood. The four boys sank to slumber, unmindful of the chorus of coyotes which visited the spot to serenade the guests of the plains.

Deep sleep wrapped the camp in silence when the promise given by the sky at sunset was fulfilled. The sleepers outside were aroused by the advance guard of the rain. In the west there arose a solid wall of ebony. Deeper and deeper grew the blackness which crept up, blotting out the sky. In the middle was a surging, rolling, tumbling mass, from the centre of which flashes of lightning shot forth. The air, which all day and night had been hot almost to suffocation, grew chill. A great calm filled the whole breadth of the plain. A hush was on earth and sky.

Then the appalling silence was broken. The huge black wave was cut by the vivid lightning, and the earth seemed shaken by the terrible battle of the elements. The muttering thunder increased in tone and volume until all space was filled with the sound. Bellowing, roaring, and crashing it came on, the clouds covering the sky and casting a mantle of blackness over the land that was only broken by the constant flashing of the lightning. The storm burst upon the camp, and with one mighty sweep almost tore the tent from its fastenings. Snow caught one end and Dyche the other, and throwing themselves flat on the ground, they held the cloth close and prevented the wind from getting under. The force of the wind crushed the tent to the earth, broke the poles, and threw the whole party flat. Then the very flood-gates of heaven seemed to be opened, and torrents of water came down. The tent was at the foot of a knoll, and down this the water swept, covering everything with drift, rubbish, and mud. Dry ravines became torrents, and the heavens seemed to send down rivers. For two long hours the storm raged, and then the turmoil ceased.

Cold and wet, the jolly crowd of yesterday lay almost perishing in the mud which flowed through the crevices under the tent. Morning dawned at last, and one by one the members of the party crawled out. Dyche had suffered most, for he had lain all night at the door of the tent with little covering, and the water and mud had poured over him, chilling him to the bone. Never before had he felt cold as he felt it then. Barely able to move, he got out at dawn, started a fire, and finally succeeded in rubbing sufficient vitality into his benumbed limbs to be sure once more that life was pleasant.

It took but a short while for the camp equipment to become dry in the sun, and by noon barely a sign of the storm was to be seen on the surrounding prairie. The remainder of this day and several days thereafter were devoted to the search for fossils. One very hot morning Snow said that he desired certain specimens of the snake kind, and wanted the boys to devote the day to collecting them.

Gathered around the camp-fire that night, all were ready to tell of their adventures and show the trophies of the chase. In handling a specimen of rattler Snow was a little careless, and the reptile slipped its head from his hand, fastening its fangs in his finger. It was a small snake, but none the less dangerous. A ligature was tied around the finger and the poison sucked from the wound. After the excitement was over the "game" was displayed. There were whip-snakes, bull-snakes, hog-nosed snakes, rattlesnakes, and racers. The oddest was the hog-nosed snake. It has the appearance and shape of the rattler without its fangs. Coiling up, it throws back its head and opens its mouth so wide that its jaws are dislocated, the result being that it is compelled to go around with the mouth wide open until the dislocation is reduced. If it bites it does no harm, for it has neither fangs nor poison-bags.

"Did you ever see a snake sunstruck?" Snow asked, and all but Mudge thought he was joking and kept a discreet silence.

"I am not quizzing. It is an actual fact that snakes cannot stand the heat of the sun on the sand. Unless they get shelter during the middle of the day they will die. If you will notice, all snakes are coiled up at noon about the foot of a bush or are under rocks. I got one big fellow to-day, a whip-snake. I had him in my net, and laid him on the ground by my side while I rested. In fifteen minutes he was dead. He was sunstruck."

The campers were in need of fresh meat, and the conversation gradually drifted to hunting. Antelope had been seen every day, in groups of from five to ten, but no attempt had been made to get one, as all the time had been devoted to collecting insects and fossils. The only fire-arm in the camp was an old carbine belonging to Mudge, and it was apparently in such a dilapidated condition that no one cared to risk his reputation by attempting a shot with it.

"Antelope is the hardest thing on earth to get a shot at, young feller," said one of a party of cowboys that visited the camp, "and you'll find that they can make out a hunter miles away. If there's only one in the band, he'll always keep his eyes skinned for something to get scared at."

This duly impressed the boys with the idea that an

antelope had almost supernatural powers, and that it would be an extremely difficult feat to shoot one, especially with Mudge's carbine. Dyche, however, announced his intention of trying to get one, and for an hour endured the gibes and jokes of the whole party. He started early next morning, and late in the evening, just as old Jim was beginning to prepare supper, he returned carrying a fifty-pound load of antelope on his back and dragging the old carbine after him. He was tired, but after a good supper of antelope steak was able to tell the story of the day's adventures as they all sat around the camp-fire.

"I tried all the forenoon to crawl near enough to the antelope to get a shot at them, but they always saw me long enough beforehand to run a mile or two away by the time I started to crawl to them. About one o'clock, a band of seven came down a ravine and went behind a knoll about three hundred yards from where I was sitting. My shoes were off in an instant and I made a rush to intercept the animals as they rounded the knoll. I made a good race, but found that the jarring of my feet on the ground had alarmed the antelope, and I got to the mouth of the ravine just in time to see the seven tawny-yellow creatures racing away about one hundred and fifty yards off. I sent two bullets after them, and in less time than it takes to tell it, those animals were going over the top of a hill fully half a mile away. They stopped on the top of the hill for a few seconds, looking back and evidently wondering what it was that was following them. My blood was up now, and I determined to follow the band. I travelled two miles, and then saw that it would be as profitable to follow the south wind.

"I knew what would be my portion if I came back to camp without meat, especially since my feet were full of cactus thorns and had been cut and bruised by the long tramp over the sand and rocks. I limped painfully back to where I had left my shoes. I went stumbling along, jumping now and then at the sound of a grasshopper, which suggested the wicked whir of a rattler's tail. As I climbed over a knoll I saw an old buck antelope standing near a dozen or more which were lying sunning themselves. The band was on the top of a hill, and the old fellow who was on guard was standing where he could see the whole country in every direction. I got within half a mile of the antelope without being seen. Two or three members of the band appeared to have their heads in the air all the time. They were evidently guards, and the safety of the band depended on their vigilance.

"Having had no experience as an antelope hunter, I was puzzled. I did not know how to get near them. Concluding that they were no harder to kill than the animals I had hunted since boyhood, I began a careful stalk. Crawling about four hundred yards up a ravine, I reached a spot within five hundred yards of the animals, unobserved. Now there was no ravine in which to crawl, so I chose the best ground I could find and began a snake-like movement up the slope. I accepted anything for cover, ant-hills, bunches of grass, cactus bushes, or even little ridges in the sand made by the wind. In this way I made two hundred yards in two hours, and had not been seen.

"Sometimes the antelope would come in such plain sight that I was compelled to lie close to the ground while they were looking my way. I got ahead at the rate of about a foot a minute until I was within two hundred and fifty yards of the old buck on the hill. By lifting my head carefully I could get glimpses of several others, but almost despaired of getting closer, and had no faith in the old carbine for such a long shot.

"While I was resting and getting possession of my faculties, the animals moved a few yards to the south. I saw, at the same time, that the ground was lower to my left and was traversed by an old buffalo trail. Moving sideways a few inches at a time and hugging the ground desperately close, I succeeded in getting to the trail, which entirely hid me from the animals. This was satisfactory to a hunter if nothing more. After resting a few minutes I pulled, dragged, and pushed myself along the trail fifty or seventy-five yards, and was now within two hundred yards of the band, and had neither been seen, smelled, nor heard. I lifted my head very slowly and carefully some six inches, and could see, through some bunches of grass which grew near my face, the entire band of fifteen or twenty animals. I could scarcely believe my eyes, but there they were in bold relief against the sky—some lying down, others standing at ease, and half a dozen or more fawns running and playing like young colts.

"I became so interested in watching the movements of the animals that I forgot for a while what I was there for. It seemed a shame to fire into the band with that old blunderbuss, but in order to get fresh meat and to put a stop to certain jokes which would be my portion if I returned to camp empty-handed, I resolved to do my best and kill an antelope. I took a long breath and trained the weapon on the nearest full-grown animal, remembering the advice which my father always gave me to 'be sure that the sights are on the gun before you pull trigger.'

"I fired, and in less than a second, apparently before the bullet had time to get there, that whole band was in motion. It was a grand sight. In their excitement they ran within fifty yards of me, and had it not been for an accumulation of rust, I might have extracted the empty shell and got another shot at them at close range. The spotless fawns had no trouble in keeping up with the band; in fact, they seemed to be playing as they ran along, for they leaped and bounded in the air as if they enjoyed the sport.

"In less than a minute they seemed a mile away, and in a half-minute more they were out of sight. They impressed me with the fact that they were the proudest, swiftest, and most graceful animals I had ever seen.

"By this time I remembered that I had shot at one of them, and with a feeling of anxiety I walked over the hill to where the band had been standing. To my great surprise I saw a big old antelope lying there dead. It almost took my breath away. I looked at the dead antelope and then at the old carbine, and decided that the old gun was the best shooter on earth. I got my shoes, and lightening the load as much as possible, put the antelope on my back and brought it in. From the feeling of my shoulders I think it weighed two hundred pounds."

It weighed about seventy, but Dyche was tired.

"Well, now, Dyche, I want you to tell me what you noticed about the animal. You have been studying about large mammals ever since I first knew you, and are too much of a naturalist not to have examined the first antelope you ever killed." This question from Snow at first startled Dyche, but after getting over a little diffidence he said:

"After my first feeling of surprise, I began to look the animal over. It was a fat, barren doe."

"A what?" said Dick. "What is a barren doe?"

"A doe that has never had a fawn, or loses her fawn early in the spring and goes without one for the rest of the summer, is called a barren doe. My attention was attracted particularly to the head, which is much more handsome than that of any of the deer kind. The muzzle was delicately and well formed. It was completely covered with hair, with the exception of a narrow streak between the nostrils and the bare skin around them, which was very black. The ears were small—smaller than those of the common deer, terminating in a point turning inward from the general direction of the ear. The eye was large and dark hazel. I had seen it described as 'black,' 'very black,' or 'intensely black,' and I made a close examination to satisfy myself. It was a hazel which, at a short distance, could easily be mistaken for black, but it was not black. The horns were small, not more than an inch long."

"What do you know about horns, Dyche?" asked Snow.

"I am coming to that. The body was short, thick, and heavy, and looked bunchy. This thick, bunchy appearance, by the way, is more noticeable on a cold day. The tail is painfully short, and the question naturally arises, 'What is it good for?' It is too small for either ornament or use, being less than three inches long. No other American animal has such trim feet. The hoofs are long, slender, and sharply pointed, being ebony black in color. One very noticeable and striking peculiarity of the feet is that there are no accessory hoofs on the back, such as are seen in cows, sheep, goats, and deer.

"Now as to the horns. The most remarkable characteristic of the antelope is that it has true horns and not antlers, and it sheds these horns every year. Naturalists have been doubtful on this subject for years; but it is a fact, nevertheless, that such is the case. All species of deer shed their antlers. There is a wide difference between horns and antlers. True horns are seen on cows, sheep, and goats, while antlers are found in the deer family. An antler is a pure bony structure growing from the frontal bone of the animal. The males of all species of deer grow their antlers every year. The antler begins to grow in the spring about the time that leaves begin to sprout on the trees, first appearing on the animal's head as a sort of knot or knob, covered with velvety skin. This knob grows very fast, soon being several inches long. It then bifurcates, and two knobs are seen on the end of the protuberance. One of these forms the main branch of the antler, while the other is the first prong or tine. The main beam continues to grow and bifurcate until the form of the antler with its many branches is completed, which occurs about the last of August. This is a wonderful growth when it is remembered that the antlers of the elk and moose sometimes attain the weight of fifty or sixty pounds. While growing, the antlers, especially the growing ends, are very tender, and can be whittled with a knife almost as easily as a green cucumber. As it approaches the base the antler becomes hard and bony in its structure. About the time the leaves of the trees have reached their growth and have ripened, the antlers have also reached their growth and are ripe, as it were. They are now very hard, and although covered with the velvet skin, it does not take long for the bucks to rub them clean by raking them up and down against the trees and bushes and whipping and threshing around in the underbrush. After the velvet has been removed this whipping and threshing process is continued until the antlers are very hard and bear a fine polish. The natural color is white, but this continued rubbing in the dirt and against the bark and leaves of trees gives them a brown color, as seen on the deer killed in the forest."

"Well, what is the use of the antlers, anyway?" asked Dick. "They get their growth and then are shed again in a few weeks."

"I think a buck without antlers would not be in style," was the reply. "The antlers are weapons of warfare with other bucks, and add much to the attractive appearance of the animal when wooing the doe. When a buck fights any other animal except another buck he usually uses his feet, but during the rutting season the bucks fight terrible battles. They tear up the earth and break down the trees in their struggles with each other. Sometimes they fight so fiercely that their horns become locked, and they die from exhaustion, becoming prey for wolves.

"The antlers are usually dropped soon after the rutting season has closed. Common deer, moose, and caribou begin to drop their antlers about the last of December, old bucks shedding them earlier than young ones. Elk usually carry their antlers until March. After the antlers drop off the skin of the head grows over the spot, and all bucks are 'muleys' until the warm spring days start the horns to sprouting again.

"Now I am ready to explain what I said about the antelope's shedding its horns. The part shed is the shell, which is a true horn and grows on the bony horn-core which is never shed. This horn is unique and interesting in several respects. About half-way up from the base is a prong or branch, which is usually rather short. It stands forward, outward, and upward. No other true horn is branched. This peculiarity has given to the animal the name of prong buck, by which it is known to many naturalists. The prong buck sheds his horns in a most remarkable way. The true horn is modified skin, as the antler is modified bone. When the time for shedding the horn comes, a new skin begins to grow between the shell and bony core. This core is similar to that of a cow and stands up a little beyond the prong. The old horn does not drop off and then let the new one grow as in the case of antlers, but the old horn stays on the core to protect the new one while it is growing. As the new horn-skin develops, a lot of coarse hairs grow from it and penetrate the old horn shell near the base, where it is quite thin. This tends to retain the old shell in place as a protection to the tender horn growing under it. As the new horn grows it produces a hard tip which fits on the end of the bony core, and this hard portion gradually extends or grows down over the bony core toward its base. When the entire horn is hard, then the old horn or shell is dropped off, leaving the bony horn-core covered with the new horn-skin. The new horn, when it first sees daylight, is a queer instrument of defence. Its upper part is a true horn, while the lower part is a thick skin covered with coarse hair. It grows harder and harder all summer just during the period when the true deer have no antlers, and is perfect by the first of August."

"Well, there's one thing about antelope," said Snow; "its meat can't be beaten."

"It's all right for a few meals," was the reply of Mudge as he took another mouthful, "but for a steady diet it is not so good as buffalo. It cloys the stomach when you get too much of it."

Dyche's success induced him to make another trial a few days later, and he went after more antelope; but late in the afternoon he returned to camp with an immense rattlesnake trailing over his shoulder. The reptile was one of the largest ever seen by any member of the party, being four feet eight inches long and proportionately thick, resembling a bull-snake. After supper Dyche told his story of the capture.

"Seeing a band of antelope on the top of some chalk bluffs, I slipped along to get a shot at them. As I went carefully over a ledge I heard something drop behind me, and looked around in time to see this big fellow coiling for another spring. He had jumped at me from a secluded place in the rocks, and missed my foot as I stepped on a projecting shelf. I went up the side of that bluff in a hurry, thinking nothing more about the antelope. I had my insect net with me, and thought he would be a fine fellow to capture alive; so I made a cast, covering him completely, much to his surprise, but I was the most surprised of the two before he got through with the net, for it was hardly a second before the snake had coiled and torn the net to pieces. He struck it viciously several times, and then began looking for me. I dropped a big flat rock on his back, which I thought would crush the life out of him, but he came out from under that rock in a hurry, mad clear through. He struck the rock several times, leaving great green splotches of venom on it. I was considerably worked up myself by this time, and began pouring such a fire of rocks upon him that I soon put him where he could do no harm.

"I tried to flag some antelope to-day, but they had been tried before. I saw a fawn in a patch of grass, and as he was a little fellow I thought I could catch him. He saw me and went out of sight like a shot, and I then heard a doe on the slope above me uttering a series of queer squealing, whistling noises. I followed and saw a herd of antelope. As they were in a position where it was impossible for me to approach them unobserved, I thought I would see what there was in flagging. Carefully adjusting my red handkerchief to my gun, I raised it in the air and waited results. As soon as the handkerchief was seen there were a number of whistles, and then the whole band trotted off and did not stop until they were a mile away."

Camp was now moved to a spot near Sheridan, on the railroad; but here it was found that eternal vigilance was the price of peace, for the whole country was overrun with tramps. They were so numerous that it was necessary for some one to remain at camp all the time to protect the supplies. This became so annoying that Dyche concluded to try the effects of a scare upon the unwelcome visitors. In the tent was a bag of live snakes, and as a dozen or more of the tramps were lounging around the camp, Dyche brought out the bag and began taking the reptiles out and allowing them to squirm about his body and head. With both hands full he started to where the vagrants were making themselves at home, and was pleased to see that his plan worked well. After that it was only necessary to begin handling the snake-bag to clear the place of tramps.

The location was found to be undesirable on account of tramps and hydrophobic skunks, and it was determined to move to Colorado, where a few weeks were spent in the Platte Cañon, completing the collection of insects.

A mule deer buck.

CHAPTER III.

In New Mexico — Encounter with a Bear — Cold Hospitality — The Retired Boston Merchant and his Mountain Ranch — An Able Missouri Liar — The Mule that was Not a Mule — Seven Deer with Seven Shots.

MONTEZUMA CAMP, on the Gallinas River, in New Mexico, which had been pre-empted for two summers in succession by Professor Snow, was an ideal spot for an insect-collector's camp. He made this his home camp in the mountains and enjoyed life with his family during the summer months in collecting and arranging the various rare insects which abound in that part of the country. Dyche was with him, but the limited sphere of work was irksome to him, and at last the desire to get among big game became so strong that he started on a tramp up the mountains with Walter Russ, a student who was on his first collecting trip. Dyche wanted to hunt deer and possibly bear, and Russ thought he could stand it as long as Dyche could.

A ranch eighteen miles from Montezuma Camp, on the Gallinas River, was the objective point for the first day, and Dyche determined to reach that place by night if it was within the range of human possibility. Under ordinary circumstances eighteen miles in a day is not a long trip, but when it is over a country where there is not even a trail, and through a cañon where half the time must be spent in getting over rocks and across the river, it will be seen that the task before the two men was anything but a pleasant one. Dyche carried a forty-two-pound pack and a gun, while Russ had a lighter load; but before they reached the end of the journey the loads felt as if each weighed a ton.

Rain began falling a short time after they started and poured down steadily all day, soaking their clothes and making the packs weigh heavier and heavier with every mile traversed. The rain so increased the volume of the stream that the slippery poles which served for foot-bridges became unsafe and the travellers were compelled to wade the river thirty-seven times, with water hip-deep. The pouring rain had so soaked the travellers, however, that the wading had little effect in increasing their discomfort. Russ was tired and ready to give up, but was cheered on by Dyche, who used all his persuasive powers to keep his companion going until they could reach the ranch.

While stumbling along, exhausted from their extraordinary exertions, they rounded a point of rocks and came face to face with a big black bear. The animal was scrambling around the ledge, evidently trying to keep his feet dry, and was within thirty feet of the two men when they saw each other. The bear rose on his hind feet in an attitude of defiance, missed his footing, rolled over into the stream with a snort of surprise, got up with a double snort of terror, and made a wild rush up the opposite side of the mountain, sending rocks and mud flying in **his** haste **to get away.** Russ made equally good time up his side of the mountain, all the while begging Dyche not to shoot but to let the bear get away if it wanted to.

Hurriedly cutting the strings which bound his gun to his pack, Dyche got the weapon and shot two big holes through the air up toward the place where the bear was making such frantic efforts to get away. The shots served only to accelerate the movements of both the bear and Russ, and in a minute the bear was out of sight, while Russ stood shivering away up on the side of the mountain. The episode had one good effect. Russ, for an hour, had been begging Dyche to stop and camp, but now he was most anxious to push on, and he hurried up the remaining two miles as if he did not know what fatigue was. **He** continued to urge more speed until they reached the park **in** which **Johnson's ranch** was situated, and only then did he feel perfectly safe.

Darkness compelled them to feel their way along, and with outstretched hands they groped about until they found a house. A knock at the door brought a man whom they asked for shelter. What was their surprise when they received a short answer that they could get no accommodations there. Not only was admission to the house refused, but even the stable was barred against them. Begging and pleading was of no avail, though the rain was pouring down on them. This aroused Dyche's indignation, and he gave the proprietor a piece of his mind about his inhospitality which had the effect of causing the man to go inside for consultation, and after this reluctant consent was given for them to enter.

The occupants of the cabin, two men, a woman, and some children, eyed the visitors with suspicion, and to tell the truth they were anything but reputable-looking citizens as they stood before the big fireplace with their clothes dripping wet and their faces and hands cut and scratched from contact with bushes and rocks on the trail. Dyche noticed that the woman seemed frightened, and to calm her fears he took from his haversack a package of coffee and from his pocket a dollar which he offered her to make him some coffee. This partially dispelled the cloud of suspicion which rested upon them, and soon the travellers were making a meal from the cold lunch which they had in their haversacks and the steaming coffee. Securing a place beside the fire they were soon sleeping the sleep of exhaustion.

Next morning Johnson, the proprietor of the ranch, appeared ashamed of his inhospitality of the previous night, and explained that it was not from any desire on his part to mistreat the travellers, but that he was not in good condition to accommodate strangers and felt particularly sensitive on the subject, as he had had a very bitter experience with a party of Englishmen a few weeks before. These men came to his ranch, asking for accommodations, and put on such lordly airs and found so much fault that Johnson and his wife determined to allow no more such people on their place. When Dyche and Russ came they supposed that it was another party of Englishmen and took measures to prevent a repetition of the insults. Johnson said he had been partly revenged on the Englishmen, as a silver-tip grizzly bear had demolished their camp which they had made about a mile from the ranch.

The Englishmen were there to get a bear and were supplied with all that money could buy in the way of hunting outfits and delicacies for the table. They were great hunters (in their own estimation) and bragged continually about the big game which they had slaughtered. Bear-tracks were plentiful on the mountain, and day after day was spent by the whole party looking for the animals. One night, on returning after a fruitless hunt, they found that the grizzly had shown his American contempt for all things British and had literally captured the camp. It hardly seemed possible that one bear could do so much damage in so short a time. Sacks of sugar had been torn open, eaten, and trampled on the ground; dried fruit was scattered over the whole place; cracker-boxes had been opened and the contents chewed and stamped in the dirt; flour-sacks were torn to pieces and the contents made into a paste and trodden into the earth; a hundred pounds of bacon was trampled, torn, and bitten until its usefulness was gone; a box of candles was chewed up and spewed over the ground; three sacks of different-sized shot had been torn open, the contents thoroughly mixed and so scattered that they were of no more value; the tobacco case was opened and several pounds of smoking and chewing tobacco strewed around and mixed with the general mess around the camp. Bruin had fully satisfied his appetite and then displayed the general "cussedness" of his nature by leaving a universal wreck behind him.

And how those Englishmen did rage! They swore vengeance against that individual bear in particular and all bears in general. There was absolutely nothing left about the place fit to eat. Going to Johnson's they begged sufficient food to last them until they could send down the mountain for more supplies. They bought a big bear-trap and imported a fine English hound and began a campaign against the bear. The greatest loss was the death of the hound, which was caused by the bite of a trade rat. One man was now constantly on guard at the camp while the others searched the woods for the bear. The only indication they ever had of his presence afterward was when he ate up a deer which one of their number killed and left in the woods while he returned to camp for assistance to bring it in.

The top of the mountain was still seven miles away, and Johnson told Dyche of another ranch there belonging to a man named Harvey, who could probably give them better hunting-grounds than would be found lower down. This determined the hunters to push on up the mountain. Near the top they came upon a beautiful park, in the centre of which stood a lone, one-roomed cabin. A cow and burro grazed on the luxuriant grass close by, and a man was chopping logs. As the two men approached he rested from his labor and looked at them with surprise. The visitors opened with a friendly "hello!"

"Hello, there, yourselves," was the response. "Where in time did **you** fellows come from?"

"From Johnson's ranch."

"I **want to** know!"

"Is this Harvey's ranch?"

"**Yes;** but how in time did you fellows get **up** here?"

"Well, we saw burro-tracks leading this way and followed them."

"I want to know! Well, you did mighty **well to** find the place."

By this time the whole party was seated on one of the logs and the equipment of the visitors fully inspected by the owner of the ranch, who finally said:

"Say, what **in** time are you fellows going to do with those fish-nets? There ain't no fish on the top **of** this mountain. I have been here going on six weeks and it ain't rained a drop since I came here."

"Where did you come from, anyway, that you don't know insect-nets?"

"Well, I want **to know!** I heard about you bug-hunters down to Vegas. I wish you would catch all the green-head **flies.** They pester my old cow and burro almost to death."

"**Y**ou don't **mean to** say that a cow and burro can live up here?"

"Live up here? Why, there's the finest grass up here you ever saw. Look at my potatoes down there; they were planted three weeks ago to-day. Did you ever see anything like it in your country? It hasn't rained a drop since I came, neither."

"**Is** there a good place to camp around here?"

"Camp? Well, I want to know! How are you going to camp?"

"Why, we have blankets and ten days' provisions in our packs."

"Camp? Yes, camp right here in the house," pointing to his one-roomed cabin.

"We have come up here to collect insects and plants and perhaps to kill animals. We did not come up to inflict ourselves on you or to bother you in any way, unless it might be to get a little milk from that cow now and then."

"We have the finest milk up here you ever saw and you can have all you want of it. That's about all we do have that's fresh, except air and water. If one of you fellows don't kill a deer pretty soon there won't be any meat on this mountain."

"Are there many deer on this mountain?"

"Deer? I want to know! Why, you can see deer 'most any day over on that hillside in the quakin' asp. You can stand right here in the door and see them pass along any morning. They've been shot at several times, though, and are gettin' kinder wild, but they're here."

"Can we camp down in that point of quaking asp?"

"No, you can camp right here at the house, cook at the big fire-place, and sleep in-doors. It gets mighty cold up here at night."

At this juncture a tall lady, whose bearing was in strange contrast with her surroundings, appeared at the door, saying:

"High, dinner is ready."

HIS MOUNTAIN RANCH.

"Come in to dinner, boys. We haven't got much. Only been up here six weeks. That's a hard mountain to pack stuff up with burros," said Harvey, and the party entered the house.

A more primitive cabin could hardly be imagined. The furniture, cut from timber, was simply and strangely made. A large fire-place at one end of the room glowed with a bed of living coals. A curtain in one corner hid a pole bedstead, while in the corner diagonally opposite a pole table, at which were several three-legged stools, completed the furniture. A lone window let in light through small plates of mica which had been quarried from the mountain-side. Long poles, overlaid with sod, formed a compact roof and kept out rain and cold. Here with his wife and two boys, Clarence and Allie, Harvey made his home.

With no apology from Mrs. Harvey, who served the dinner daintily, all sat down to a meal composed of baked beans, Boston brown bread, and coffee. It is unnecessary to say that the tired tramps made the most of it. During the meal it transpired that Harvey was an ex-Boston merchant, whose good nature had run away with his judgment to such an extent that he finally found all his capital and profits trusted to those who either could not or would not pay, and he was compelled to seek a new life in the West. In his wanderings he reached this spot, and his eyes were so filled with the beauty of the place that he determined to make it his home. Water and grass were there, while trees grew on the hillside. What more did he want? He built his cabin and prepared to stand the heavy snows and low temperature which the ranchmen of the valley told him he would experience. To all their tales of woe his invariable answer was:

"I want to know!" *

Dyche and Russ accepted the invitation which was given with such sincerity, and several days were spent searching for rare insects and plants. When a goodly number of specimens had been secured Russ returned to Camp Montezuma with them and Dyche remained at the ranch intending to hunt animals and birds. Harvey's continued talk of the number and apparent tameness of deer gave Dyche a desire to obtain a few for specimens and at the same time for a supply of fresh meat at the ranch. With this double object in view he started every morning early and returned every evening with monotonous regularity with the same story to tell. He had seen deer, but just as he got ready to shoot they disappeared in the woods and he could not find them again. At last he went out, saying he would not come back until he got a deer. He left the ranch early with a single biscuit in his pocket for lunch. It was almost dark when they heard three wild whoops away up on the side of the mountain. An-

* Four years later Harvey's labor had borne good fruit and his mountain home, ten thousand feet above the level of the sea, had become a favorite resort of tourists. The ranch is the highest on the American continent and is peculiarly isolated. Fifty acres are now under cultivation and over a hundred head of horses, cattle, and burros graze on the grass of the meadow. Tons of hay, barley, wheat, rye, and oats are raised, and from his dairy he sells many pounds of butter and cheese. Notwithstanding the altitude, the climate is salubrious, and the several log-cabins which he has built find tenants each summer. During the winter he finds that the temperature is rarely below zero, while the snow-fall is light, and the valley is suitable for occupancy the entire year.

swering shouts **rang** out, and **the ranch** was immediately a scene of excitement **and** bustle, for the signal was understood. Dyche had got a deer.

"Hurry up there, boys, and get that pack-saddle on old Reuben. Don't you hear that feller hollerin'? He's got a deer, sure!" called out Harvey.

Away they hurried up the mountain, and at the edge of a bunch of quaking asp they found Dyche standing **beside** a fine young buck, while at a little distance lay **a** second equally **fine. The deer** were placed on the burro and the procession went to the ranch, where they were met by Mrs. Harvey. Everybody was talking at once and no one cared what the other had to say. **The hour was** late and one deer was dressed and quartered. Harvey showed that he was a good cook himself, and while his wife prepared the biscuits and coffee he broiled the venison steaks over the coals in the fire-place. The hot supper was enjoyed by all, and then Dyche told the story of his day's hunt.

"I had walked **all** day without getting sight **of** anything worth shooting. I was tired and disgusted and was dragging myself home when I reached the bunch **of quaking** asp and willows, up there. I had **been along** there several times before but had always gone around the thicket. This time I went through it. I got down on my hands and knees and crawled in. The ground **was wet and** mossy, but that made no difference. After proceeding about a quarter of a mile in this way I reached a small grassy park in the centre of the thicket. Just as I entered it on one side I saw three deer going out on the other. They stopped for a minute at the edge of the timber and my Winchester, which was already pointed that way, was fired at a fine young buck. I took as good an aim as was possible in my nervous state. The buck bounded into the forest with the others, and in a second I saw one of the number stop an instant so that I could get a glimpse of his body as he passed between the trees. I shot a second time, taking even more care than at first. The smoke came back in my face and made me very uncomfortable. I walked slowly and carefully across the opening, watching for the slightest movement, but all was still and I could neither see nor hear the deer. I found plenty of tracks and then great spots of blood. Now I knew my chances were good, I went rapidly forward on the trail and a few rods further found the buck lying by the side of a log where he had made his last leap. He was stone dead. I felt like cheering, but restrained myself and began searching for more signs of deer. I thought I might have wounded another and did not like to leave a wounded deer. I followed one trail for about a quarter of a mile, but saw no signs of blood and then returned to the other track. This led me only a few yards and there I found the second deer, shot dead in his tracks. Now was the time to yell, and yell I did, and you know the rest of the story, for you have eaten fresh meat."

"I want to know! And this is the finest meat I have ever tasted, too."

Several days later Dyche again went out with the intention of getting a deer before he returned. Away up on the side of the mountain he was making his way carefully, stepping softly and avoiding stones and twigs in his pathway for fear the noise would disturb the game. At the opening of a beautiful mountain meadow he stood entranced at the picture which was presented to his gaze. At his feet, off to the west, lay "Paradise Valley," as Harvey had named his ranch. Down through a narrow defile in the mountains, as through a golden frame tinged with green, could be seen a picture the like of which had never been limned by painter's brush. Looking over the beautiful valley, all aglow with the beams of the morning's sun, the circling crags of the highest mountain held in their mighty grasp, in peaceful seclusion, the ideal home of one tired of the world and its constant warfare. Out through the cañon, up which the winding trail led, was seen a vast cleft at the foot of which lay a mirror-like lake, reflecting the beauties of hill and dale. The grassy slope led off on either side, and the clusters of fir, spruce, and quaking asp trees formed a living picture in vivid colors.

While looking at the scene and breathing the pure air which gave him new life, Dyche came upon a large doe. The startled animal was hurrying across the open ground and had just reached the edge of the timber, when the gun spoke and she dropped in her tracks with a ball through the shoulders. The viscera were removed and the blood wiped out with grass and leaves, and the ranch was reached just as Harvey was preparing to get out of bed. At the word "deer" the ranchman was half out, and before the word "killed" was uttered he was dressing.

"I want to know! Another deer! Clare, get old Reuben and bring him up here in a hurry. We'll go right after that deer. Rhea," calling to his wife, "Dyche has another deer. Don't it beat time! Say, Rhea, give us an extra mess of that venison this morning. Well, I want to know! If he didn't bring the liver with him! Cook a good lot of that too. How far is it? Just think of it, Rhea, just up here at the head of the meadow. Three deer killed almost within a quarter of a mile of the house. This is the country for me. They wouldn't believe it if I should write it back to Boston."

Clare brought up old Reuben, and while Mrs. Harvey cooked the venison the deer was brought down from the mountain. The old doe was a good load for the burro and much trouble was experienced in making the contrary brute carry it down safely. How that breakfast was enjoyed! The long morning tramp and the bracing air sharpened the appetite, while the savory venison, flanked on either side by hot gems and washed down by such coffee and cream as can only be obtained at Harvey's ranch, made a meal not easily forgotten.

Dyche now had two young deer, buck and doe, and an old doe for specimens, but the mountains had too great a charm for him and he could not leave them. The early morning hunts and the trailing of the deer through the woods were too great a pleasure to be soon foregone. A few days after killing the doe a rain fell just at the dinner-hour and continued for a few minutes after. When it cleared and the sun came out Dyche took his Winchester and started up the side of the mountain. Trees and grass were heavy with water and he was soon soaking wet. While going carefully through the underbrush, making as little noise as possible, he was suddenly startled by a noise about twenty-five yards in front of him. A deer jumped out and ran with long bounds for about fifty yards, when it stopped and looked back. Just a small portion of the back was visible through the underbrush. Taking careful aim he fired, and making his way through the bushes found that he had broken the back of a doe. Her hair was in fine condition and in every way she was a splendid specimen. The old signal-yell was given, and answering shouts came up from the valley as Harvey and the boys with Reuben hastened to get the quarry.

As Harvey had occasion to go to Las Vegas, Dyche sent a good load of venison to the party at Camp Montezuma. At Las Vegas Harvey picked up a man who said he was a Missourian and brought him along to work for the summer. According to the Missourian's story, he was the greatest deer-hunter that ever lived. He had killed twenty-five the winter before. Twenty-four of them were shot through the heart. One of them, a big buck, had been out of sight except the head, and its neck was broken just back of the ears. So many stories like these were told by the Missourian that Dyche and Harvey took him out on the mountain to hunt deer. Passing down a meadow with the Missourian about two hundred yards ahead, two deer suddenly jumped up some forty yards in front of the deer-slayer. They stood looking at him and he returned the stare. At last they got disgusted and moved slowly off into the woods. Harvey and Dyche rushed down to the spot and breathlessly asked:

"Why in thunder didn't you shoot at those deer?"

"What deer?"

"What deer! Well, I want to know! Didn't you see them two deer at the edge of the brush?"

"I didn't see any deer."

"Well, I want to know! I guess you had it bad. They looked you right in the face and waited for you to shoot. You'll never get another such chance."

The alleged deer-slayer was the target for so many jokes about the affair that he never again mentioned his hunting qualifications.

Dyche now concluded to extend his field of operations, as he had hunted thoroughly over the ground in the vicinity of the ranch. With this end in view he went about four miles over the mountain into another natural park. Mexicans had built a slight inclosure and kept their animals during the summer up in the park. There were at the time about a dozen horses and mules in the inclosure. The Mexicans rarely visited the place, and it was as wild as any part of the country. While following the fresh trail of a deer, Dyche suddenly came upon an animal standing in the brush. After a long examination he thought that it was a mule. Going a little closer he made another examination and was almost sure it was a mule. He walked slowly nearer and could see part of the side and one leg. He was now very sure it was a mule and walked boldly forward to within thirty yards of the animal, which suddenly gave a great bound and went crashing through the woods, making a noise like a herd of cattle. In a very few minutes it was out of sight, and Dyche realized that he had missed an opportunity of getting a fine bull elk. He followed it seven or eight miles, but overtaking that elk was out of the question, so the hunter returned to the ranch and spent several days meditating upon his loss.

An old Virginia buck deer ranged through the mountains and had been seen several times; but he had been hunted, and it has not yet been chronicled if there is anything wilder or more tricky than a Virginia deer that has been hunted. This old fellow in particular seemed to be equal to all the snares and traps laid for him. One cool, frosty September morning Dyche carefully made his way through the woods frequented by this particular deer. While skirting along the edge of a piece of timber he espied a moving object across the opening. It was about half a mile away, but he could see that it was a buck. It had its head high in the air and was on the lookout for danger. His general direction lay down a ridge. Dyche crawled to a point where he calculated the animal would pass. It was cold, and this with his nervousness put Dyche in a tremble. Minute after minute went by and no deer came. He was just about to give it up, when the deer appeared in a bunch of quaking asp about eighty yards away. He had deserted the ridge and was hiding in the thick woods. Hardly daring to breathe, Dyche trained the gun toward the old fellow, when something attracted the animal's attention and it gave three or four great bounds with its head high in the air and stopped, stone still, listening. Then the rifle spoke and the buck crashed through the woods at full speed, but he was jumping high. Soon he came to the ground in a heap, and the signal-yell brought the boys and Reuben and the finest specimen of the trip was taken to the ranch. With seven deer killed with seven shots, Dyche was now ready to return to his duties away from the woods.

As he appeared in the opening.

CHAPTER IV.

In New Mexico—Hunting Bear—Once more at Harvey's—The Same Luck—Camp Bear Trail—The last Pot-Shot—A Long Chase—In Ambush—Too Much of a Good Thing—A Monarch Dethroned—What was Done with the Royal Hide.

T was under greatly changed conditions that Dyche climbed the trail leading to Harvey's ranch, a year after the incidents related in the last chapter. He now knew what to expect in the way of game and went well prepared for the summer's campaign. He had given the subject much thought during the winter months, and this time he was after bears. For his companion he had a student who displayed nerve, endurance, and tact—three indispensable attributes of a good camper and collector.

It was early in June when Dyche and Brown left Las Vegas and took the old trail up the Gallinas River. The day was fine and travelling good, and they reached Harvey's home in good season. Dyche was little prepared for the change which had been wrought in the mountain ranch. Harvey had utilized every moment, and now in place of the little one-roomed cabin there stood an imposing log-house with shingled roof, a log-barn partly up, and the ex-Boston merchant had prepared five acres each for oats and wheat, while two acres were in potatoes. His herd of one cow and burro had increased until he had several of each kind. Substantial furniture replaced the pole table, bedstead, and stools, and to crown all there was a large cooking-range in the kitchen. As all these additions had been transported up the narrow trail on the backs of burros, it was always a mystery to Dyche how the stove had been brought in. The two old-fashioned fire-places gave to the whole a home-like appearance which brought many pleasant memories to Dyche.

As the naturalists neared the ranch they saw Harvey at work in his field, and Dyche gave the signal-yell which had been used last year to notify the ranch of the death of a deer. As the sound reached Harvey he gave a jump and sent back the answering yell, while from the house rushed the boys, and even old Reuben recognized the sound and added his sweet voice to the general clamor. They came pell-mell down the trail, with Reuben bringing up the rear; and Dyche was soon surrounded by the trio, each trying to shake his hand and all talking at once.

"There's not been a deer killed since you left here," said Harvey. "Beaty and his partner have a cabin up here above the meadow, where they are mining mica. They have been hunting deer and hunting deer all winter and got nothing. I guess they have either run them all off or made them so wild that you won't be able to get one. I've been telling them how you got one at every shot and whenever you wanted one."

"Well, we'll have something better this year. I've

got a new gun, a regular bear-gun. It's the old reliable Sharp's rifle."

"I want to know!"

"Yes. We are going to have bear-meat this year."

"I want to know! I saw bear-tracks down below the meadow not over a week ago, and not a month ago one ate up my calf that died. There are bear here. Beaty and his partner have been after them, but they get nothing."

Dinner was soon ready and all went into the house to attend to the inner man. Reuben was a privileged character, but Mrs. Harvey let him understand that his privileges ended at the door of the house, and he stood near the step chewing an old dish-rag which had inadvertently been left outside. The most noticeable change about the place was on the table. Plates, knives, forks, spoons, and dishes were there, and they were set on a fine extension table. But when the viands were spread there was something which could not have been changed—that freshness and flavor of the wilderness which brought back such a flood of memories to Dyche that he could scarcely finish the meal, so anxious was he to get out again among the quaking asp and fir.

After dinner Brown went back down the trail with a train of burros to get some articles which had been left at the foot of the mountain. Dyche unstrapped his new rifle and wandered off over the old familiar hills. He cared little for the game he might find. His whole being was filled with the joy which comes only to those who have roamed over the fastnesses of the mountains in perfect freedom—a thing of which the naturalist hunter alone can boast.

Just as the sun was going over the western range the sharp crack of a rifle was heard in a willow patch up on the side of the mountain, and this was soon followed by the familiar shout. When the gunshot was heard Harvey could hardly wait for the signal, and he yelled to the boys to saddle old Reuben, for he was sure Dyche had a deer. When the call came ringing down the hillside pandemonium reigned at the ranch. Harvey alternately sang, danced, and shouted, and then scolded the boys for being so slow.

"Get on the skillet, Rhea—Dyche has a deer. I knew he would get one. Hurry up there, boys. What in time are you poking along so for?"

Harvey, the boys, and Reuben were soon at the clump of trees and found Dyche standing by the side of a fine doe. It was carried to the ranch in triumph, Harvey alternately congratulating Dyche and then himself on the success of the hunt. He was impatient to let Beaty know of it, and told Dyche that he would not have had that deer missed for anything. He would rather have given fifty, yes, a hundred dollars than to have missed getting that deer.

"I knew you couldn't help getting one. How in time did you find him so easy? Beaty never could see one in this part of the mountain."

"I spent the afternoon wandering around and thinking of the fine times I had last year. I recalled how I had stalked a buck there and had been disappointed under that tree; how a doe had once jumped out of that clump of bushes and how I had killed a deer in this opening. I paid little attention to anything, but just wanted to breathe the fresh air and smell the fir trees. Just at sundown I reached the edge of that clump of willows and stalked carefully through it, as it seemed a likely place to find game. Away across on the other side I saw something move. Watching carefully I soon made out that it was a deer, and stood perfectly still, while the animal walked out into the opening and began nibbling grass. It was a long way to shoot, over two hundred yards, but I knew if I moved the animal would see me and be off like a shot. I set the sights on 'Old Reliable' at three hundred yards, took the best aim possible, and pulled trigger. When the smoke cleared away I went over and found the old doe with her back broken."

But Dyche was after bear this year and had no time to spend at the ranch. Three days later, with a string of burros hired from Harvey, the naturalists started over an old trail to a point higher in the mountains, where man had rarely been. A chapter might be written on the haps and mishaps of that trip. The trail was almost obliterated and the hunters were unused to carrying packs, and the trials and tribulations of that journey were quite enough to discourage a man of less persistence. It might be told how the packs would slip, and how the donkeys would go the wrong way at the wrong time; how they would refuse to cross streams at critical moments, and how one obstinate old jenny had deliberately broken through the crust of snow and almost frozen her legs before they could tunnel her out of the deep drift. But bears the party must have now. They were encamped in a little meadow which marked a pass between two snow-capped ranges. Each night on the way up the mountain had been full of incident and adventure. It is a story in itself to tell of the terrible wind which blew the dead trees down about the camp during the first night on the trail; how the monarchs of the forest came crashing down in all directions and threatened destruction to the campers; how the tent was crushed and the donkeys almost frightened out of their wits by the storm; how the following night some wild animal, probably a bear or lion, had scared the burros so that they rushed into camp and would not venture out again after grass; how old Reuben improved the opportunity to make a meal of the dish-rag and had eaten half a bar of soap before he was discovered and tied to a tree with a very short rope in order that he might meditate on his sins. All things must come to an end, and this trip up the mountains was no exception to the rule.

Just at dark on the evening of the third day camp was made in a little mountain meadow, and when daylight came next morning they saw that the place could not have been better chosen. Within seventy-five yards of the tent was a well-worn bear trail, where the brutes had passed from one mountain to the other for years. The place was christened "Camp Bear Trail," and preparations for an extended hunt were made.

There was no doubt that there were plenty of bears in the vicinity, for their trails could be seen all about the place. So confident were Dyche and Brown that they would **soon get** a bear **that they** contented themselves with bacon rather than risk scaring away the big game by a shot at a grouse or **deer.** Every night the burros were brought into camp **and tied** for fear they would fall a prey to bruin. In a few days the novelty of the situation wore off and Dyche killed a deer, not caring whether **the shot scared** a bear or not. **He went out on the side of the** mountain to look for meat, carrying his big Sharp and a number 10 Colt shot-gun so that **he would be ready** for anything. **Hearing a noise in a** clump of willows his blood almost boiled with excitement, for from the noise he was sure that a bear was coming. Suddenly the bushes parted and a big mule deer buck trotted out into the opening with head up and ready to jump at the slightest sound. He was on the steep hillside fifty yards above Dyche, who sent a load of buckshot into him and the animal rolled down **to** within thirty feet of his slayer.

Bear-trails were followed day after day **without** ever seeing or hearing anything of the animals. An ambush was laid for a deer which came to the little lake every evening for water. While lying in wait a band of the animals came down to the water's edge to drink. When they were **well** bunched and not over seventy-five yards away, Dyche discharged " Old Reliable," and **as** the frightened deer ran he fired three more shots after them. When the smoke cleared away he found that he had done that which no true hunter feels proud of. He had allowed his excitement to get the better of his judgment, and there, as the result of his four shots, lay six fine deer.

Four had seemingly been shot through at the first fire. Dyche was ashamed and almost vowed to give up hunting on account of the unwarranted slaughter. He made all the amends in his power, and every pound of meat and all the skins were saved. Brown went down the trail to Harvey's with the meat. In his whole hunting experience thereafter this served as a good lesson. He never again shot at random into a bunch of animals, but always singled out the one wanted for meat or a specimen.

The first night after Brown went down with the meat a big bear passed along the trail, leaving a track as big as a peck measure. Dyche resolved to see where that bear went if it took all summer. With a light lunch in his pocket, a thin rubber blanket, some matches, a hatchet, his rifle, and seventeen cartridges he started on the campaign. The trail was fresh and the bear did not seem to know that he was followed. He went swinging along, leaving a trail that could be followed on the run. Here he had turned over a log and there he had scratched up the earth looking for roots and tender shoots. He wandered around in an apparently aimless manner, and Dyche followed every track. Here a stream had been crossed and the water was still muddy where the big fellow had stopped to wallow. The trail led into a deep fir forest and it was almost dusk under the trees. The pines interlaced at the top and the ground was covered with a thick bed of needles, shredded fir-cones that had been opened by squirrels looking for the seeds, and leaves, which formed a carpet in some places three feet thick. In this mass of débris were found many bear beds, where the animals had scooped out great hollows and made comfortable sleeping-places. Night settled down and Dyche was at last compelled to give up the chase. He had travelled fully fifteen miles over the mountains and through the forest, and was so tired that he did not think of returning to camp, but finding a sheltered place on a large projecting rock, he spread his gossamer blanket as a wind-break, and on a bed of spruce and fir boughs, with a blazing fire at his feet, he tried to sleep. He was tired, but sleep would not come. He would fall into a doze and then wake with a start from dreaming about a fight with a big grizzly, and would alternately doze and build up the fire until morning came. At early dawn he started back to the home camp, and the day was spent resting.

Awaking from a refreshing sleep next morning, Dyche saw where a bear had come over the trail, and, to show his contempt for the hunters, had wallowed in the spring from which they got their drinking-water. Brown returned with mail and provisions from Harvey's and a council of war was held. It was decided to make an ambush and wait until the bear again went over the trail and then kill him. He evidently passed along in the night, and a platform was built in a tree near the trail.

Darkness found Dyche safely ensconced on the platform, prepared for an all-night's siege. A heavy overcoat was supplemented by a blanket and a trunk-strap secured him to the tree in case he should fall asleep. The rifle and shot-gun were within easy reach, and it would have been a bad night for a bear had one come across the trail. All night long, shivering and longing for daylight, Dyche sat there, but never a sound of bear was heard. The stillness was horrible. Not an owl hooted and not a twig was snapped by fox or wolf. The twittering of the early birds at last announced the approach of day, and Dyche crawled down, cold and benumbed, and made his way to camp, where a hot breakfast soon reinvigourated him. Again and still again was this ambush laid. A band of deer went over the trail, and then a fox came and smelled the bait but did not touch it, and a wildcat came along and clawed around, but went on without going near the bait. Dyche let them all go, as he did not want to shoot and run the risk of scaring away a bear. But no bear came. Soon after sunrise on the third morning Dyche crawled into camp for a little breakfast and then hastened back. The bear had been there during his absence. The old fellow evidently came along a few seconds after Dyche left, for he had eaten the whole of the bait which had been left near the trail, and then had scratched up the earth near by. To nish the performance he had wallowed in the little stream and passed on over the mountain.

Dyche was tired, sleepy, sore, and stiff, but this was too much for human endurance and he promptly started on the broad trail left by the animal. The bear went along, turning over logs, stones, and stumps, looking for bugs. Here he wallowed in a mud-hole to relieve himself from fleas, and there he scratched up the earth or stretched himself up on a tree. Dyche could not catch up with him, and at two o'clock in the afternoon he returned to camp almost worn out. A big buck jumped from cover, but he let the animal go. A light supper, and a blank until dawn. A hot breakfast restored him, and after a whole day spent about the camp he felt like going on another campaign. A fox and a wildcat were caught in traps and the skins taken care of, and then Dyche started out to look at some traps, expecting to be gone about an hour. It was late in the afternoon when he returned, and from the flush in his face and his general excitement Brown knew that he had seen bears. Dyche had had a most wonderful adventure, and was so wild over it that he could hardly wait until after supper to tell his story.

"I got to the fox-traps, and as I was looking around I saw a large bear-trail that was very fresh. The bears had been here, there, and everywhere. The ground was dug up as if a drove of hogs had been rooting and overturning the logs and stones. There must have been a herd of them, for paths led through the woods in a dozen different directions. I took a large circuit in order to find which way they had gone. I soon found the main trail, which was as easily followed as if a herd of cattle had been along there. It went through a number of grassy parks, down a small stream, up another, and then over a mountain. I followed as rapidly as possible, expecting every minute to see them. They spread destruction in their path. Logs, stumps, and stones were turned over and ant-hills torn to pieces. A choke-cherry patch was stripped of berries and leaves. The bushes were torn and stripped and the tops chewed off, presenting a sorrowful sight. I determined to get back to camp and start to-morrow with five days' rations in my haversack, and find those bears or die in the attempt.

"I wandered along revolving my plans in my mind, and came out of the big woods on a mesa about two hundred yards wide, flanked on one side by a heavy forest, while on the other was a sheer fall of several hundred feet. It was a beautiful place, and I thought it would be an amusing occupation to roll stones down the cañon, but was too tired to put the idea into practice. I was walking slowly along, looking now and then towards the woods, but not thinking of seeing anything, when suddenly there appeared at the edge of the timber a number of moving objects. I could not make out what they were, but there was such a number of them I concluded they must be goats. Mexicans sometimes bring goats up the Pecos River into this country, and I thought a herd might have been driven out of the regular trail. As they appeared to be coming towards me I waited and soon got another glimpse of them about three hundred yards away. They were among the trees, and the sun through the leaves gave them a spotted appearance which convinced me that they were goats, for many of the Mexican goats are spotted. I could not see the herder and stood perfectly still waiting for them to get nearer. Suddenly there came out of the forest, directly to the west of me and not over seventy yards away, a huge grizzly bear.

"Before I could realise what had happened, out came another, then a third, a fourth, a fifth, a sixth, and a seventh. Just think of it, seven big bears in sight all at once! I think there were four more which I saw, making eleven in all in that band. I knew I was in a most desperate situation. On one hand was a bottomless precipice and on the other a herd of the most ferocious animals which range the mountains. How the sweat did roll off of my face! There was only one thing to do, and I did it to perfection. That was to stand perfectly still and let those bears go about their business. I was hunting bears, but not these particular bears. There I stood in perfectly plain view of those animals, but they did not see me. They were walking fast, and I had a splendid opportunity to observe their mode of travel as they passed on.

"I no longer wondered at my not being able to overtake them on the trail. They went swinging along in a sort of shambling trot or canter almost as fast as the gait of a horse. Some would stop for a second at a time, turning over logs and stones, and then hurry on to overtake the band, which moved right along.

"As soon as they were out of sight in the woods to the southwest of me, I hastened to assure myself that I was still alive and wiped the sweat from my face. I could easily have put a bullet through any of them, but what would have happened then? I might have been set upon by the whole gang and would not have made a fair meal for one of them. I made haste to get into the woods and tried to head them off. I wanted to get a shot at them where I could get shelter in the trees if they attacked me. They unintentionally outwitted me, however, and went up a ridge while I was watching a stream."

"Well, I have a scheme," said Brown. "Let me go back to Harvey's and then down to Las Vegas and get a big bear-trap, and we will get a bear, sure. It will take some time, but it seems that we are spending more time than anything else, anyway."

"Well, as we need more provisions I guess you'd better make the trip to-morrow," was the reply, and so it was settled. Brown got an early start next morning and Dyche was left alone. The trip would take about ten days to go and return. Dyche started for the woods to see if he could get another look at his bear herd.

On the evening of the ninth day after Brown left, Dyche heard the song of a burro down the cañon and he hastened to get a warm supper for Brown, who he knew was coming. From the back of old Reuben dangled a fifty-pound bear-trap. Hot coffee, biscuit, and broiled steak were soon smoking on the table, and Brown went ravenously to work on them.

"This venison is mighty tough," he remarked. "It must have been one of the oldest bucks in the mountains."

"Take another piece," said Dyche.

"It will be better after we've had it a week or two," was Brown's comment, as he took an extra tough bite. "What kind of meat is this, anyhow? It's the toughest venison I ever tasted."

"Maybe it's fox."

"Fox nothing. It's more like burro-meat, I should say. I didn't leave any of the jacks here when I went away, did I?"

Dyche could keep his story no longer and burst out with: "It's bear-meat, man. A regular old grizzly at that."

"What? got a bear! Well, if this is a piece of him it must have been the one old Noah had in the ark. Well, I'm glad he didn't get you. Where's the skin? How did you get him?"

"The day you left camp I started out to look at that big trail where my herd went along. I thought there might be some satisfaction in looking at the track if I couldn't see the bears. The trail was a day old, but I followed along without exactly knowing why. After following it for miles I started back to camp, and reached a grassy slope on the side of the mountain and sat down to rest in the edge of it. There was a willow patch in front, and to the east of me and across from the willows was an almost impenetrable forest of spruce trees. Flowing through an opening in this forest was a little stream which joined another rivulet flowing from the willows. As I sat on a log looking across this stream at the spruce forest I saw something moving among the trees, and from the glimpse I got of it among the spruce branches I thought it was a deer. I watched very carefully, expecting to see a big mule buck step out into the opening.

"To my great astonishment a huge grizzly bear stepped from the forest at the opening made by the little stream. What a monster he was! He must have been as big as a cow. The wind was in his favour, and getting scent of me he placed his front feet on a log and began sniffing the air. I could see his big head going up and down, and must confess that I felt a little chill run over me. The old Sharp's rifle always seemed so big and heavy before, but now I wished it was a cannon. I took the best aim possible, holding my breath to prevent muscular movement, and remembering the advice of my father to always see that the sights were on the gun before pulling trigger, then I fired. The gun belched forth its load with a roar which was echoed by another roar from the bear.

"Here he came growling, rolling, tumbling, falling, jumping, and bellowing, making a terrific noise. I slipped off my shoes, reloaded the gun, placed a handful of cartridges in the crown of my hat by my side, and waited. I thought the whole gang might appear and wanted to be ready for any emergency. The old fellow came on towards me, and I determined that if he ever crossed that stream I would give him another 520-grain bullet. He would get tangled up in a fallen spruce tree and would tear himself loose in a most wonderful manner. Now he was in the willows, rolling and tumbling and biting everything that was in his way.

"His strength and activity were simply wonderful. One blow of his mighty paw would have killed the greatest prize-fighter that ever lived. I have heard stories of men killing grizzly bears with their knives, but I don't think it possible for twenty men to have stood before that bear in his death-agony. I could now see him very plainly, and could see that he was covered with blood and was getting weaker and weaker every minute as he came on towards me. Just as he reached the edge of the water he spread himself out on all-fours, and there continued throwing up his head, uttering most horrible groans and guttural grunts, while I sat cold and spell-bound under the great excitement. At last he died, seventeen minutes after he had received a ball which would have been instant death to an ox. Then I got up and went over to where he lay.

"He was a monster indeed. Not fat but so muscular. Streams of blood were running from his mouth where he had broken his great teeth in his death-agony. I was under intense excitement, but I noticed that his legs were black while his sides and back were of a tawny tint. His tail was very short, so short, in fact, that he could not even sit down upon it.

"It had been raining all day, but I never noticed it as I sat on the log watching the dying throes of the bear. I must confess that I had a pang of remorse as I looked down at the dead monster. I had at last outwitted one of the giants of the forest, but in his death I had seen the qualities of a grand warrior. After finishing my examination of the big fellow I turned about and went to camp, leaving him just where he had fallen. I reached the camp at dark, and would have given a good deal if you had been here to share the enthusiasm with me.

"There was no sleep for me that night. I went over that fearful struggle again and again, and when I dozed off I would wake with a start from a frightful dream of the bear. Next morning I was rested but not refreshed, and after a hurried breakfast I hastened down the cañon where I had left the dead bear. It seemed at times as if it might all be a dream—but no, when I got to the spot there he lay, just as I had left him the night before, dead and cold. Having spent about two hours in taking seventy measurements for future reference, I skinned him. I found that the old fellow had been shot before, for there were two bullets about the size of a forty-four Winchester imbedded in his body, one in his hip and the other in the shoulder. My ball hit him fairly in the neck, cutting the jugular vein and passed entirely through the body, coming out about six inches from the tail near the spine.

"I was almost worn out, but I carried the meat, skin, and head to the big snow-drift and buried them, and dragged myself to camp, where I ate a light supper and then rolled up in my blankets and slept until dawn next morning."

During the succeeding days Dyche thoroughly dressed the skin. All fat and flesh were removed and the feet skinned down to the very toe-nails, and all ligaments removed from the bones. A preparation of one part alum and four parts salt was now rubbed all over the skin. The feet and head were folded in and then saturated with a strong solution of the mixture. This operation was repeated in twenty-four hours and then again in twenty-four, and the skin was ready to be hung up to dry.

"They are sure good bear knives."

CHAPTER V.

Making a New Trail—The Storm on the Mountain-Top—
Neither Bear nor Elk—A Journey in Utter Darkness—
Catching Foxes—Unwelcome Visitors—Brown's Avalanche—The Bear was Handicapped—An Experience with Ants and Fleas.

AFTER several days of fruitless search for more bears, it was decided that the animals had left that particular part of the mountains, and the naturalists concluded to move camp. Cacheing the surplus meat in a snow-drift and packing the camp equipment on the backs of the burros, the little animals were headed towards the north star, and the party went through the woods, making a new trail. They did not have the least idea of where they were going, except that they wanted better hunting-grounds, where they might possibly get bear or elk. It was said in that vicinity that elk had been seen on the ridge between the Pecos and Canadian rivers, and this ridge was made the objective point. Their way led through heavy spruce timber which lined the ridge, and about noon they reached an open mesa on the side of the mountain, which had been burned over. For this reason travelling was difficult and they were compelled to cut their way.

While working through this down timber, two big

mule bucks jumped up and started to run, going over and under the logs with remarkable ease. Getting sight of one of the bucks as it was going between the trees about three hundred yards away, Dyche fired, but with no apparent result. Getting another opportunity he shot again. The burros were left and search made in all directions for the deer; but nothing could be found, and Dyche was just about to give up when Brown, who had gone on about seventy yards farther, shouted:

"Here! here! here's your old buck, dead as a doornail."

Pushing on up the mountain, camp was made at dark by a spring on the west slope. This place received the name of Camp Wild Buck, but, owing to later developments, was named Hungry Cañon. Cacheing the greater portion of the venison in a snow-drift the next morning, they travelled three or four miles north until the ridge was reached. This was the watershed between the Rio Grande on the south and the Arkansas on the north, and led towards the Jicorilla Mountains, gradually rising and becoming narrower as it extended above timber-line.

At one o'clock they reached a high point in the country, and from this place was seen one of the grandest sights in the high ranges. Off to the west was a vast ocean of mountain-tops, the timbered slopes being surrounded by high peaks, running above timber-line.

The river, winding in the cañon far below, was a mere thread of silver among the trees, thousands of feet down. Through a narrow opening of the mountains was a view of the plains far to the east, which, with the rough, rock-girt frame, presented a picture as of the field of the cloth of gold. The purple hills made a rich, dark foreground, while the golden sunlight flecked the prairie with the colour of living fire.

So entranced with the scene were they that the flight of time was forgotten, and when they returned to mundane things they realised that if they wished to camp where there was grass and water, they would have to be finding a way down to timber-line. Spending an hour in this effort, they again glanced at the beautiful scene; but now all was changed. A mass of clouds, through which lightning played, and which grew denser and denser as it came, was rolling up the valley of the Pecos. Flash after flash could be seen, and then faintly came the sound of thunder.

The sight was interesting, and as they watched it their attention was called to the other side of the range by an answering reverberation. Looking down the Canadian valley, to the north, they saw another storm-cloud rolling towards the ridge upon which they stood. As the storms approached nearer and nearer they discovered that they would soon be in the midst of a battle of the elements.

The sky became overcast and it grew dark. The play of lightning became fiercer and fiercer on both sides, and the peals of thunder soon merged into one continuous roar. Peal was answered by peal, and the echoes took up the terrible sound and sent it on and on until all space seemed filled with the tumultuous noise. There came a gust of wind, and then for a second a noiseless calm was over all. Then the two clouds rushed to the top of the intervening ridge, and the space between them seemed filled with one continuous sheet of flame. The whole mountain was the battle-ground and heavenly fire the weapons in this awful contest of the elements. Lightning flashed through the air in all directions. Rocks rolled down the side of the mountain, while a hail-storm sent missiles over the valley. The burros huddled together and trembled at the raging of the elements, while Dyche and Brown threw themselves flat on the ground and sought the shelter afforded by the jutting rocks.

Gradually the uproar ceased. The hail continued to fall until the whole mountain-top was covered to a depth of two or three inches. It was some minutes before the storm, now broken into numerous smaller ones, left the mountain and permitted the naturalists to start down to timber and shelter.

Within an hour from the time the storm began, the whole country was smiling as if nothing had ever occurred to mar its serenity. Not a breath of wind ruffled the trees. Far down the mountain appeared the little lake, no longer vexed by the wind, and picturing on its breast the wooded hills. A pale blue haze hung over everything, and covered the landscape with a veil which, like distance, lent enchantment to the view.

A few days spent here proved that there was neither bear nor elk in this part of the country, and camp was moved, the party going in an easterly direction and travelling all day through a wilderness of down timber until four o'clock in the afternoon, when a little park was reached. The whole space in the park was covered with grass and beautiful flowers, and, darting here and there through the opening, were what appeared to be tiny balls of fire. At first Brown and Dyche were completely mystified, but an examination showed these fire-balls to be hundreds of rufous humming-birds darting among the blossoms. They whirled and flashed through the air, and the whirring of their wings made music like that of wind among the trees. Camp Humming-Bird, the place was called. After a day's rest it was determined to go back to the other side of the range and see what could be found in the Canadian valley.

The top of the range was reached by noon. The place where the dreadful storm had raged a few days before was now peaceful. The atmosphere was so clear that the distant mountains, seventy-five and a hundred miles away, seemed to lie within easy walking distance. At nightfall a fine, perfectly wild, iron-gray horse, which had evidently broken away from the confinement of civilisation, was seen in the timber. This incident was sufficient to give the name of Wild Horse Camp to the spot, the tent being pitched by the side of a large snow-drift.

Late next afternoon Dyche went for a little hunt. As he passed through a willow park two bucks jumped and ran through a clump of trees. He took a flying shot as they disappeared, and as only one came out on the other side, Dyche was sure he had not made a miss. The buck jumped up on the side of the hill, and stood as if waiting for its companion.

Dyche trained the rifle across a log and fired at the deer, fully three hundred yards away. The buck made several awkward jumps, but did not run. Again and again he shot, but each time the deer would jump a little, but would not leave. The seventh shot brought the animal to its knees, and Dyche made his way to it. He found that one, perhaps the first, ball had taken off the animal's horns, and so confused it that it did not run. One ball had creased the withers, another cut the brisket, while the last had broken both the fore-legs.

Brown, hearing so much firing, put the pack-saddles on all the burros, for he had never known Dyche to shoot more than once or twice, and supposed that he had a large supply of game. The camp was in a sort of horseshoe inlet in the mountains, and this peculiar formation caused the echoes to come from many different directions. The result was that Brown started off almost in the opposite direction from the spot where Dyche had done his shooting. Tired of waiting, Dyche started to carry the deer, and had proceeded some distance before Brown reached him. Night came on rapidly, and a dense fog or cloud settled down on the mountain. They were over a mile from camp, and it was dark and raining. Dyche wanted to leave the deer and find the tent, but Brown objected. He said he wanted venison for supper, and would have that deer in camp if it took all night. The darkness was so intense that the donkeys would not travel, and Brown was compelled to go ahead while Dyche followed in the rear, holding to a burro's tail to keep from getting lost. On they went, stumbling and falling, the deer slipping from the pack-saddle and causing no end of trouble, until the whole party was almost exhausted. The deer fell off so often that Dyche insisted that it should be left, but Brown was obstinate and took it on his shoulders, saying that he must have venison for supper, and was not going to let a little thing like that get the best of him. They finally reached the opening where the tent had been placed, but they had no idea where it was, and the rain was falling in such torrents that they could not see their hands before their eyes.

"You stay with the donkeys, and I will find the tent," said Brown; and leaving Dyche he started off through the darkness. Ever and anon he called to keep his bearings, and then his calls became faint, and at last he was out of hearing. Dyche tried to start a fire, but everything was soaking wet, and all his efforts were fruitless. Brown returned after a time and announced that he had found the tent. They travelled about an hour, and then Brown began to wonder what was the matter. He had placed a lighted candle upright on the floor, so that the light could be seen. He made another excursion and found the tent, discovering that the candle had fallen down in the mud and the light had gone out. It was late, but a roaring fire soon dried and warmed their clothing, while a supper on the venison refreshed the inner man; and at three o'clock in the morning they rolled into bed and slept the sleep which follows exhaustion.

A whole day's rest next day, Sunday, prepared

them for further exertions. The day following, Dyche was working with his gun, trying to extract an obstinate shell, when Brown came hurrying into the tent with his eyes dilated and his whole frame shaking with excitement. He pointed to a bunch of trees near by, where Dyche saw a magnificent buck, about seventy-five yards away. Dropping on one knee he fired just as the animal was disappearing amid the timber. The buck was found dead a hundred yards from the place where he was shot.

It was evident that in this part of the country there was little game which they wanted, and next day the skins and venison were packed on the donkeys and the back trail taken. At Camp Wild Horse a stop was made, and Brown took the donkeys and with all the spare meat started for Harvey's ranch, leaving Dyche alone in the mountains. To while away the time during the seven days of Brown's absence, Dyche amused himself with trap-setting. He saw some martens catching conies in the rocks and succeeded in getting two of them. He carried the big bear-trap down into the cañon and set it, and went every morning and evening to see it. A V-shaped pen of logs was made and the trap set in this. Foxes were cunning enough to climb over the logs and eat the bait. To prevent this the top of the pen was covered with brush, and then the foxes dug holes under the logs. Dyche now had a time of it to outwit the little animals. He set traps around the logs, but the foxes evaded them. Then he set traps away from the pen and had the satisfaction of catching the robbers.

Six days had passed in this way, and on the morning of the seventh Dyche had started for the bear-trap, when he saw Brown coming up the cañon.

"How did you get here so early in the morning?" was the greeting.

"I got within half a mile of camp last night before sundown, but clouds and fog settled down so that I did not like to risk coming on in the dark."

Brown had the mail and plenty of provisions, which were badly needed, for Dyche had had only two biscuits since Brown left.

"I had visitors while you were away," said Dyche, after they had eaten breakfast. "You remember that camp-fire that we saw down in the valley? Well, I think the fellows who built that fire came up to see me, and from their appearance and actions I think they meant no good. I saw them coming and made it a point to be cleaning the guns when they came up. They talked Spanish, but when they found that I could not understand it they talked good English and asked me many questions about my business and why I was up here. I told them the truth, showed my specimens, and convinced them that I was all right, and at last they went away. I was not at all satisfied with the interview, and when they had gone I started to hunt and skirted along the edge of the woods where I could watch them. They went down into the valley and met another man who was taking care of the horses. I think they were a band of horse-thieves, and I lay behind a log all that night, and if they had come fooling around they would have had a reception which they did not bargain for."

A fine white-tailed buck which had been eluding

Dyche for a week fell victim to his gun that afternoon, and next morning Brown took the shot-gun and went down the cañon. Dyche was looking after his fox-traps, and had just found one in which there was a beautiful brown fox when he heard a report down the cañon as if from a cannon. Brown had evidently touched off "Old Vesuvius," as the shot-gun had been christened by the cowboys. Dyche gave a signal but got no answer. Darkness came, but no Brown. Becoming much alarmed lest some accident had befallen his companion, Dyche hurried down the cañon, calling now and then but getting no response. About two hundred yards from camp he heard a slight noise and called out. A muffled answer was returned, and then he was sure Brown had been badly hurt. Hastening down he found Brown staggering up the mountain with a big doe on his back. His face was black and blue and his clothing covered with hair and blood. A deep scratch ran along the side of his nose, and taken altogether he was a most dilapidated-looking object. He had fired both barrels of the gun at once, and, being unused to such a heavy charge, had been kicked over by the heavily loaded weapon.

Next day Brown took his insect-net and Dyche his rifle and they wandered off looking for what they could find. Dyche was first to return and had supper ready when Brown reached camp. Brown had a good story to tell and began while they were still eating:

"Say, Prof, do you remember that old crag of loose rock we passed as we came into this cañon?

BROWN'S AVALANCHE.

Well, I was looking for insects around in that vicinity, but could not find many. After fooling around trying to kill some grouse with rocks, I got one by hitting it on the head. I climbed up on that crag. There was a big rocky mass which was split off from the rest, and I got a pole and tried to pry it off. I worked for nearly two hours, for I wanted to see it go down the mountain. At last it got loose, and how it did go! I wish you could have seen it. The big rock started a hundred others and this hundred a thousand more. You would have thought the whole top of the mountain had fallen. It frightened me as I thought there might possibly have been someone at the bottom of the cañon. A cloud of dust and smoke arose which had the smell of the infernal regions, while the noise was terrific."

When he finished his story Dyche, who had been listening quietly, said he had had an adventure also.

"When I left camp I went north and then circled east and worked along the edge of the cañon. Seeing about a dozen big mule bucks in a band, I started to stalk them. Working very slowly and crawling carefully, I got within two hundred yards of them. There was a bunch of scrubby spruce trees about seventy-five yards to the east of them on the edge of the cañon, and I determined to get to that clump, where I could have my pick of the band. Leaving my hat and shoes behind, I worked my way along and got almost there. Twenty-five yards more and the trees would be reached. I stopped to get my breath a little, when a regular earthquake seemed to break loose. An immense crag away up on the side of the mountain broke from its place and came down into the cañon. The first stone started hundreds of others, and these hundreds started thousands more, and they came down with a terrible noise. Dust and smoke arose and a smell as of the infernal regions came from the bottom. When it was over the deer were gone and I came back to camp."

As Dyche proceeded with his story and ended in almost the very words of Brown, the latter stopped eating, his jaw dropped, and when the story was finished his only comment was:

"Well, I'll be darned!"

Satisfying themselves that no bears were to be found in this part of the mountain, they started for old Camp Bear Trail next morning, and late in the evening they pitched their tent in the familiar spot. The bear-trap was again set in the cañon, but several days passed without special incident. Dyche and Brown made a long trip to the west and returned late in the evening, tired from their exertions. While Brown took the burros to water, Dyche set about getting supper, and had it well under way when Brown rushed into camp, calling:

"What is that roaring in the cañon?"

An answer was unnecessary, for the sound was so plain that both shouted at once:

"A bear! a bear!"

Brown seized the shot-gun without waiting to see whether it was loaded or not, while Dyche got the rifle and a handful of cartridges, and away they went down the cañon. The roaring continued and grew louder and louder, and the two men went over bushes, logs, stumps, and even small trees in their haste. Brown fell over a bush and lost the shot-gun, but was in too great a hurry to pick it up, and on he went, shouting at the top of his voice. The noise was closer now, and appeared to be a cross between the roar of an African lion and the bellow of an enraged bull. Then the bear came in sight. He was going down the cañon as fast as the big fifty-pound trap fastened to his leg would let him. The long chain was fastened to a twenty-five-foot pole, and this caught in the rocks and bushes, detaining bruin in his frantic efforts to get away from his pursuers.

"Shoot him, Professor, shoot him!" yelled Brown.

At the sound of his voice the bear stopped and looked back. He now tried to get the trap from his foot by beating it with his paws and biting it until his mouth was bloody and his teeth broken.

A ball from the rifle knocked the bear down, but he was up in an instant and was going down the cañon faster than ever. Another ball sent him to grass again, but he would not stay down, and then a third ball knocked him over so hard that he could not get up. He now went through a series of wild contortions, rolling and tumbling, roaring and bellowing in a most terrible manner. He had received his death-wound and Dyche let him alone to die, which took about fifteen minutes. Then the naturalists shook hands, executed their favourite war-dance, and did numerous seemingly foolish things with which all hunters can sympathise. Their victim was measured and skinned and left where he fell until morning, for no animal would touch a dead bear.

The excitement of the bear-hunt made them forget for the time being how tired they were, but now that it was all over they went slowly back to get their supper. In the fight with the bear they had travelled nearly two miles down the cañon, and were almost exhausted when camp was reached. In their haste to get to the bear Brown had forgotten to tie up old Reuben, and that animal utilised the opportunity by doing up the camp while they were doing up the bear. He first made a raid on the little pole table made between the trees and cleaned it off. He ate the apple-sauce and licked the dish. The last half-pound of butter followed this. The soap was gone. A piece of bacon had been pulled down and mouthed until it was useless. The dish-cloth had been chewed, and when found the old reprobate had just finished the batch of hot bread which had been left baking in the Dutch oven covered with coals. Reuben had poked the cover off without burning himself and finished the bread. Such incidents are amusing to look back upon or to tell about, but just at that time it was the straw which broke the hearts of the naturalists. Reuben's head was tied close to a tree, where he was left to repent of his sins, which he undoubtedly did, for next morning he made a light luncheon off of one of Brown's socks which had been hung on a bush to dry.

That afternoon a voice was heard calling away down the cañon. Wondering who it could be, Dyche and Brown went down to render aid if aid was needed. It was Clare, who had made the trip up into those wild mountains alone. He shared Brown's bed on one side of the tent, while Dyche occupied the other. About midnight the sleepers were aroused by an appalling din, and Clare bolted through the tent entrance without waiting to untie the strings. He stood by the fire yelling at the top of his voice.

"What's the matter with you? Have you got the nightmare?" asked Brown.

"Ants! ants in my pants!" yelled Clare, as he rubbed his legs and tried to dislodge the insects. He was scolded into returning to bed, and the camp had barely got settled down again when the racket broke out once more, but this time it was Brown, and he was immediately followed by Clare.

"Ants! ants!" they screamed. Dyche grumbled at them for disturbing him and asked why they could not make less noise.

"It's Brown that's got the nightmare," said Clare. "It isn't so funny now." Just then Dyche felt as if a piece had been bitten out of his leg, and he gave a jump and shout that told the others that they were not alone in their misery. The rest of the night was spent in searching for the voracious insects and driving them out. The tent had been pitched near the site of a big ant-hill, but it was supposed that all the insects had been cleared out. At breakfast in the morning Brown suddenly dropped his plate and seized his leg with an exclamation.

"I've got him, sure, this time," he said, and an investigation revealed no ant, but a big reddish flea. The whole thing was plain now. There were no ants, but the bear-skin, which had been deposited near the tent, was full of fleas, and when the skin got cold the insects had made an excursion in search of warmth and food, which they found under the blankets in the tent.

That evening there was a grand feast in the camp with bear-meat, brisket of deer, and good appetites. Then came a general bear talk. Bear-meat was tough and stringy and jumped under the teeth like a piece of india-rubber. Some was roasted and some was boiled, but it was all tough and had a peculiar flavor.

"Is all bear-meat as tough as this?" asked Brown.

"No. I've eaten the meat of black bear that was very nice and tender," answered the professor. "It tasted like pork. I presume a young grizzly would not be bad eating."

"How big do bears get?" asked Clare.

"Well, that's hard to tell. According to some reports they occasionally exceed two thousand pounds, but I don't think there are many so large. From the best information I am able to obtain I don't believe they get larger than twelve hundred pounds. I was told by an old hunter that he saw a bear in California that weighed between eleven and twelve hundred pounds and it was a monster. It was kept in captivity and had been fed every day until it was very fat."

"How large do black bears get?"

"Ordinarily between two hundred and three hundred and fifty pounds. I heard of one taken in Idaho which weighed four hundred and twenty-five pounds."

BEAR TALK.

"How many kinds of bear are there in this country?" asked Clare.

"Only two, the grizzly and the black."

"Well, how about the silver-tip and the cinnamon?"

"I was just going to say that the cinnamon is a variety of the black, while the silver-tip is a smaller form of the grizzly. Bears vary greatly in size, even when full grown. They also vary much in colour, ranging from a light yellowish-brown to almost jet-black. I have seen the skins of young grizzlies almost as light-coloured as wolves. The feet and lower part of the legs are, however, dark, shading from black to various tints of brown. The colour also depends on the season. In the spring the old shaggy hair is much lighter than the new fall coat. This wide range in colour and size of the grizzly has done much to encourage the notion that there are different kinds of grizzlies in the United States. The cinnamon or brown bear seems to be only a colour variety of the black. Those who claim that these bears are different species say that the black and cinnamon cross and make the different colours. This is really an argument against the claimants, for different species of animals do not cross. If they did they would merge finally into one single species. The silver-tip seems to be a well-marked variety of the grizzly."

"How could one tell a silver-tip grizzly from a cinnamon?" asked Brown.

"A grizzly can usually be distinguished by the fact that his front claws are twice as long as his hind ones. They sometimes reach the length of five inches. In the black or cinnamon there is not such a marked difference. The colour of black or brown bears is more uniform throughout, while the colour of the grizzly is variegated. The grizzly also has a more distinct mane, which stands up between the shoulders, sometimes reaching a length of five or six inches. The feet of the grizzly are always broad and thick as compared with those of the black bear."

"How large are bears when they are born?"

"About the size of Norway rats. The old bear generally has twins and sometimes triplets. An Indian gave me some young ones which he said were about two weeks old. They were about the size of Newfoundland puppies."

"Do bears hibernate?" asked Brown.

"Yes, after a fashion. Not in the ordinary sense of the word. They do not get torpid. They usually 'den up' in the colder parts of the winter. These dens are simply nests of leaves and grass under the roots of some overturned tree. Here the bear stays until the weather gets warm, unless he is disturbed, and then he is 'up and coming.'"

"Do bears voluntarily attack people?" asked Clare.

"If you suddenly came upon an old female with cubs she would probably show fight just as a cow moose, an old sow, or any domestic animal would. A wounded bear will also fight just the same as any other wounded animal."

"You don't mean to say that all those stories about bears attacking people are false, do you?" asked Brown.

"Most of them are undoubtedly so. Bears are the most cunning animals in the mountains. I have come upon their beds while they were still warm, but the bears were gone without my even getting a glimpse of them. No matter how sly I am, they are pretty sure to hear me and go scurrying through the woods as fast as their legs will carry them. Their senses of hearing, smelling, and seeing seem to be marvellously developed."

"What do they feed on?"

"That depends to a considerable extent upon the country in which they live. Like the hog, they take anything they can get. They will take meat, rob birds' nests and suck the eggs, catch frogs, eat fish; they are very fond of vegetables, berries, and tender roots, and they will eat ants and other insects and even worms. They are very fond of acorns and go a long way after them. In Washington and British Columbia they desert the mountains when the salmon are running and live along the streams."

The hunt was now over and the skins and meat were packed to Harvey's ranch, where the naturalists said farewell to their friends and returned to civilisation.

CHAPTER VI.

A Peculiar Wedding-Journey—Field Taxidermy—A Typical Mountaineer—Big Bear Talk—The Cabin was Looted—A Lost Timepiece—A Great Day for Talking.

ALL was bustle in Paradise Valley one May morning two years after the events narrated in the preceding chapter. Harvey was reading to Mrs. Harvey and Allie a letter, the gist of which was as follows:

"Meet me with the burros. I have two companions and am coming to spend the summer with you."

It was from Dyche, and this announcement was the cause of the commotion. Everyone was impatient for the pack-train to be off, even old Reuben seeming to know that something out of the ordinary had happened, for instead of hanging back and causing trouble when the train started, he walked sedately to his place and waited for his saddle.

Dyche and his companions had travelled the twenty-five miles from Las Vegas to the foot of the mountains in a wagon, and were now waiting for the arrival of the pack-train from the mountain ranch. Their baggage had been dumped on the ground at the edge of a little Mexican village on the banks of the Gallinas River, and here they made their camp. The home life of the Mexicans, as seen from the door of their tent, served to while away the time of waiting. The fact first observed was that every man and woman in the village was an inveterate cigarette-smoker. The children had driven their herds of goats to the mountains and the lazy oxen were drawing their creaking and groaning carts or hauling heavy logs to the sawmill, which puffed and buzzed a short distance down the river. Fires were started in the big out-door beehive oven ready for the week's baking, and village life was in full blast, when the campers were startled by a loud hail in a voice that to Dyche had a familiar ring.

"Ha-o, there! Como le va?"

"Como le va yourself."

"Who's this you've got with you?"

"This is my wife. I'm married now," replied Dyche.

"Well, I want to know! So you're married! And you thought you would come to the finest country in the world for your wedding-trip?"

"I see that you haven't lost old Reuben yet."

"Lost him? Why, you couldn't lose him if you tried. He's just the meanest old burro in the whole country. He steals eggs and eats young chickens every chance he gets. If you and Brown had killed him for a specimen when you were here, it would have saved me much trouble and bad temper. I had to bring him along, for I darsn't leave him at the ranch."

"Well, why can't Mrs. Dyche ride him to the ranch? He's gentle enough and she will be in no danger in going over the trail."

No sooner suggested than acted on, and the saddle

was put on the old burro, and for the next six miles Mrs. Dyche exercised her arms trying to make him keep in the trail and up with the other donkeys. He had a mind of his own and would wander off to lunch on tempting bits of grass away from the trail. Her feeble efforts at punishment had about as much effect on him as tickling with a straw. Finally he was put in the middle of the train and better time was made. The ranch was reached by noon, and by night the party had become thoroughly domiciled, ready for the summer's campaign.

The object of this year's trip was to put into practical operation an idea of Dyche's regarding field mounting of birds. The camp was placed by the side of a cool spring which gushed from the edge of a patch of quaking asps, while on the west was a heavy body of spruce. Stretching to the south for two or three hundred yards was a fine grassy meadow, merging into a jungle of willow and tag-alder bushes covering a piece of swampy ground. A small wall-tent was devoted to taxidermic operations, and soon became known as the museum. This tent faced south, overlooking the meadow, and was prepared for any emergency. Shelves arranged in one corner held the mounting outfit, which included a bunch of wire, a can of alum and arsenic, a bottle of benzine, a can of plaster-of-Paris, a case of the size and shape of a cigar-box containing a complete assortment of tools for skinning and mounting birds, a roll of manilla paper, and a few other articles. Under these shelves were kept rolls of cotton-batting and excelsior. A pole table served as an operating-desk, while a hollow log, sawed to the proper length and covered with buckskin, furnished the seat. On one side was a pole platform on which mounted birds were stored, while a similar one held the guns and ammunition ready for instant action. By noon on the second day the whole place was complete, and they were ready for anything of the bird kind from an eagle to a "hummer."

Two beautiful long-crested jays perched themselves on a limb in front of the tent and began an inspection which cost them their lives, and they were first to be immolated on the shrine of science. The advantage of field taxidermy was shown in the first day's work. The operator was not compelled to rely upon either memory or notes to ascertain the correct positions of birds, but from his desk could look into the trees and see the counterpart of the one in hand and thus get the natural poses and positions. This practice soon convinced Dyche that the live bird in its natural habitat was the only safe guide to be followed. Another point gained by field-work was in mounting many of the smaller birds which had such tender skins that it was next to impossible to carry them long distances before mounting. This was especially noticeable in the hummers, which, every taxidermist knows, are exceedingly difficult to mount after becoming dry. Ordinarily many skins are spoiled by oil oozing from the shot-holes, but when they are mounted immediately this can be obviated with little difficulty. Doves and pigeons, which are so difficult to mount from dry specimens, were handled very readily, and the fresh skins were found to be tough enough fcr all practical purposes.

The month of June, which was thus devoted to field-work, passed very quickly, and one hundred and fifteen birds stood on the pole table ready for transportation down the mountain. But this was a problem which had not yet been solved. How were the mounted specimens to be carried down the twisting trail and over the hundreds of miles of railroad without injuring them? Dyche had an idea about it before he came, and several cracker and soap boxes had been taken apart and carried up the mountain and were now put together. The birds had been mounted on a T-shaped stand, and the cross-piece of the T was now taken off and placed on the bottom of the box, holes in which permitted the passage of wire. This wire passed over the cross-piece and was securely fastened below. Adjusting the birds at such an angle as to occupy the least space, a layer covered the bottom of the box and then cleats were screwed on and a shelf or tray, also covered with birds, was placed in, the process being continued until the box was filled. The lid was screwed on and the whole neatly covered with canvas, giving to it the appearance of an ordinary satchel, to which handles were attached. In addition to the hundred and fifteen mounted birds a large cracker-box was filled with dry skins, the larger ones being rolled in cotton and fastened in the box so that they could not crowd each other, and the smaller ones being placed in cylinders made of heavy manilla paper, to which they were secured by long pins passing through the paper into the **bird-skin.** These boxes were carried over trails and **on the cars** as hand-baggage. Eames, the student who was with Dyche this year, fashioned a pack-saddle for **his** shoulders, and marched ahead of the pack-animals with the two boxes of mounted birds. Arriving at their destination, it was found that the journey of several hundred miles had done no damage to the frailest specimen.

Dyche **concluded to give his wife a taste of real mountain life, and just** as he was considering a **trip to** Camp **Bear Trail, Beaty,** the mica-miner, came **up from Las Vegas on his** way to his ranch, **which was** established **at the** head-waters of the **Pecos, about ten miles from the** ridge on which Dyche and **Brown** had passed through the terrible electrical storm. The miner gave them such a hearty invitation to accompany him home that they accepted, and he promised the best hunting and fishing in the country. The hardships of a three-days' journey over the roughest part of the mountains **did not** deter Mrs. Dyche, and early one morning the **start** was made, **old** Reuben carrying Mrs. Dyche. **With** Beaty **in the lead** picking out the trail and Dyche in **the rear to** punch up stragglers, they went up the mountains, Beaty beguiling **the** way with many quaint stories.

All signs of **a trail** finally faded away and merged into **a** tangled network of underbrush and fallen timber. Dyche offered Beaty a small hand axe with which **to cut his** way, **but** the latter declined, **and** drawing an immense knife from his belt, remarked: "**This is** sure the thing for that kind **of** work.

This is sure a good knife. I made it myself out of a drill, and I made this one too," drawing its mate from his belt. The blades were twelve inches long and of finely tempered steel. "They're sure bear knives, and long enough to reach a bear's heart. If a bear ever comes across this old man he will sure feel this knife in his heart. See these buckhorn handles? I sure made them and killed the buck that wore the horns."

Beaty was a typical mountaineer, and as he stood flourishing the big knives above his big sombrero, with his buckskin coat, he looked a fit match for any bear that walked the range. His continued talk of what he would "sure" do and Dyche's knowledge of the power of the grizzly bear, made the latter a little dubious as to the outcome of a fight with Beaty and a bear as chief actors, but he kept his counsel and drew the mountaineer out until the woods were filled with the sound of his big bear talk.

The first day passed pleasantly, barring the many "unpleasantnesses" between Reuben and his rider, and camp was made by the side of a spring in a grassy meadow. The second day took the party through a long stretch of burned timber; the donkeys caused much trouble by continually running into snags and tearing their packs. Camp was made on the slope of the mountain, near a bunch of quaking asp and spruce trees. As the train approached the spot an old hen grouse flew up, and while the supper was being prepared the young grouse could be heard "peeping" in the grass and bushes. Mrs. Dyche could not rest until the little fellows had been caught and snuggled under a blanket, where they spent the night cosily; and next morning when they all ran about, alive and spry, Dyche did not regret having spent an hour on his hands and knees in search of them the night before. The mother grouse flew down from a neighbouring tree as the train moved away, and Mrs. Dyche's heart was made glad with the knowledge that the grouse family was reunited.

"Do you see that pile of stones?" asked Beaty, pointing to a heap of small rocks which had evidently been thrown together for some purpose. "A few years ago it was sure very dry on the plains, and Mexican herders brought their sheep up into the mountains where they could get grass and water. One old man had $10 in his pocket and he was killed and robbed by some of his companions. When his friends carried his body out, wherever they stopped to rest they would make a little pile of stones, and now whenever a Mexican passes along he adds a stone to the pile and says a prayer for the rest of the murdered man's soul."

"Is that the reason there are so many little stone piles throughout New Mexico?" asked Dyche.

"No, not always. They are made to commemorate some incident. They've sure got some queer superstitions, and one of their religious ones is the queerest."

"What's that?"

"A lot of them go up on Hermit's Peak, over there, and crawl around on their hands and knees among the cactus bushes and on the stones, exposing themselves to hunger and thirst. They sure die from it sometimes."

"Why do they call it Hermit's Peak?"

"An old Mexican hermit used to live there, and when the people went there they took him things to eat. They say he had a gold-mine up there and he sure had lots of gold. He hid it all away when he left, and it has never been found. There's sure gold there, but I've not found it yet. I've got colours, though."

So Beaty beguiled the way with his stories, and as they were going down into a grassy valley he suddenly stopped the train and called out:

"See that willow thicket there? Well, I sure saw a bear there. You see that old log there? Well, when I got there I heard a mighty noise in those willows and four bears came out all at once. I thought they had it in for the old man, but I just threw myself down by the side of that big log, jerked out both knives, and if the bears had come I sure would have done some of them. When they saw that I was sure ready for them they got out of there in a hurry. As I lay there four bears went hustling up the other side of the mountain. This is sure a great country for bears."

Late in the afternoon, when within a few rods of the cabin of the miner, the train was again called to a halt, and Beaty, pointing to a large tree, said:

"See that tree there? I met an old grizzly there. I'd been fishing and was coming to the ranch. I had thirty fish on a stringer, not one less than twelve inches long, and was within thirty feet of a bear before I saw him. I skinned up that tree in a hurry, and that bear came along and ate up every one of those fish and then licked the stringer. He then looked at me and walked off about his business as if nothing had happened."

"Why didn't you kill him with your big bear knives?" asked Dyche.

"I had left my knives at the cabin or there would have been the worst bear-fight ever seen in these mountains."

"He would have killed you, Beaty, before you could have hurt him with a knife."

"He would have had some fun doin' it. I would have thrown myself down on my back. That is sure the way to fight a bear. When he came up I would have plunged both knives into him. It would sure have fixed him too."

All this bear talk had a depressing effect upon Mrs. Dyche, and as the party moved on towards the cabin they saw that a bear had recently been there. The little garden had been torn up and the big tracks could be seen everywhere.

"See, there are bears here. The woods are full of them."

Unlocking the door of the little two-roomed cabin, Beaty stepped inside, but at once bounced out with his eyes distended and his whole face showing great excitement.

"I've sure been robbed. Somebody's taken everything I had, even my gun and pistol. There ain't enough left for one meal. I'll follow them. If I ever strike their trail they'll be mine."

He raved and howled until the air seemed to take on a pale blue tinge and smelled sulphurous. Mrs. Dyche concluded it would be well to go down and watch the water run in the river while Dyche tried in vain to calm the enraged miner. The cabin had been completely looted, to all appearances. Beaty pointed to places where he had had a ham or a side of bacon hanging or where a can of syrup stood, but these places were all vacant now. He took a wide circuit around the place to find a trail, but all in vain. The only tracks to be seen were those made by Beaty himself during his last visit. Down by the river were seen small tracks, like those of a woman, and Beaty came in full of the idea that he had discovered the whole plot.

"There was a white man with a Mexican woman and they had a train of burros, for they could never have carried off all that stuff on their backs. I will sure find them if it takes my whole life."

All night long the visitors could hear their host grumbling, and ever and anon he would break out with oaths that would make Mrs. Dyche shudder. Beaty was up early and started off down the Pecos River to his nearest neighbour, twenty-five miles away, in hopes of getting some trace of the desperadoes. Dyche and his wife concluded that it would be best to get out of the country themselves, and notwithstanding they had just had a hard three-days' trip up the mountains, they started back for Harvey's ranch. They had plenty of provisions and the river was full of fish, but the talk about bears and the experience with robbers were too much for Mrs.

Dyche. The first night on the way down some wild animal frightened the donkeys so that they rushed into camp for protection and kept the travellers awake the balance of the night. Next day they left the main trail and went off to Bear Trail Camp. It seemed like getting home again to Dyche. Two years had passed since the last visit, but everything was just as he had left it. An old dish-rag still hung on a peg in the tree, while on the pole table sat the wooden bowl, carved out with so much patience by Brown. Cans and tin pans were hanging in their accustomed places, while even the firewood which had been placed in a dry nook was still there ready for the fire. Sticking from the roots of the tree was the neck of a bottle of syrup, now greatly improved with age. The two were soon comfortable, and after five days' continuous travelling on a rough trail the rest was welcome.

Dyche knew every foot of ground around the camp as well as he did his father's pasture, and he started out to find a deer. Mrs. Dyche had seen so many bear signs that she insisted that he should not go out of hearing. There were so many evidences in the great holes where bears had wallowed, or where they had turned over the logs and stones, or scratched the trees, that she was sure she saw a bear in every blackened stump on the mountain. Late in the afternoon Dyche was sitting at the edge of a little meadow, concealed by a bunch of willows.

Just as the sun touched the western mountain-tops a deer came to the edge of the woods about three hundred yards from where Dyche was sitting. The animal was very cautious and stepped along carefully, watching for the least sign of danger. He finally stepped from the timber and began nipping the leaves and grass and feeding towards where the hunter was concealed. He fed along so slowly that it became a race between the deer and the sun. At last the sun disappeared and darkness came on rapidly. Dyche was uneasy lest his wife should drop a pan or make some other noise which would startle the deer. It became so dark that the sights on the rifle were not distinguishable, and when the deer came within fifty yards Dyche fired his shot-gun. The animal gave several great jumps and Dyche did not wait to see if his shot had been fatal, but he sent a ball from the rifle after it and the deer sank to the ground. Calling Mrs. Dyche, who took charge of the guns, he dressed the deer and carried it to camp. A hot supper of venison liver, biscuits, coffee, and syrup convinced them that they had chosen well in coming to the mountains. Two days were spent at Camp Bear Trail, and they lived as happily as if they had been domiciled at some summer hotel.

When they had left Harvey's ranch, eight days before, the place was comparatively quiet, with only Harvey, his wife, and Allie there. As they approached on their return, however, they were made aware of the fact that a change had come since their departure. Voices, loud and strong, could be heard long before they reached the clearing. At the house they found several mountaineers sitting around, "swapping lies" about their varied adventures. There was Fly, the mica-miner, he of the strong

A GREAT DAY FOR TALKING. 89

frame and equally strong lungs. He was a talker, and the tones were of such a pitch and volume that the tops of the highest hills echoed them back. Then there was Levette, called for short "Cockeye." Levette wanted to talk and was continually trying to get in a word, but he was entirely too slow for the crowd he was now with. The inevitable "I was just goin' to say" never got beyond his tongue's end until some readier talker broke in, and it was never known what he was going to say. Eames, who had never been defeated in a talking-match, was there, and then there was Harvey, who had had forty-nine years' experience at it. All had something to say, and none cared what the others had done; so the result was that all talked at once. Dyche was somewhat of a talker himself, and when he got into the crowd he was anxious to tell of his experience at Beaty's cabin. The result of all this was that when Mrs. Harvey announced that the haunch of venison was sufficiently baked and that dinner was ready, there were five men all talking at the top of their voices, each sandwiching in his story wherever he got an opportunity. While Mrs. Harvey and Mrs. Dyche stood in the door of the kitchen laughing at the exertions of the talkers, in walked Beaty.

Now Beaty carried off the palm as a talker in that region, and when he began all others were compelled from the sheer force of necessity to stop. Beaty had a story to tell and he told it. His voice soon made itself felt in the general pandemonium, and the others gradually quieted down until he had the floor to himself. But he talked under difficulties. He would hardly get started before someone would break in, and then all would go at it again.

"You remember when I left you, Professor. Well, it was just eight o'clock in the morning, and at just two o'clock in the afternoon I was at the ranch, twenty-five miles down the Pecos."

Here Fly broke in: "That was no walking. In California I walked from Elk Creek down to Gold Gulch, forty-two miles, in just eight hours and——"

Here someone else broke in, and the story of some great walk had to be told by each one, and Beaty was compelled to wait until the flood passed and then started fresh again.

"At the ranch I found my old partner, Everhart, and he told me that he heard I was sick down to Vegas. He sure sent a boy up to the ranch to look after things until I could get back. The boy stayed there about a week and then a big bear came around the place and would not go away, although he shot at it through the window. The boy was scared, and dug a hole under the floor and buried everything in the cabin in the hole. He took all the dirt down to the river and threw it in and smoothed the place over. Sometimes he wore a pair of my old shoes and made big tracks around the place."

"Well, I want to know!"

"By the great wars!"

"I was just goin' to say——"

"I thought you could have smelled it."

"You couldn't smell anything but brimstone around there when Beaty got there."

"That boy knew what he was about."

A GREAT DAY FOR TALKING. 91

"He sure fooled the old man."

And so the changes were rung on all the exclamations the crowd could think of. The talk continued on this subject until dinner was over, and then as a fine, drizzling rain was falling, conversation drifted into other channels, while the men crowded around the big fireplace. Harvey started the ball by laughing at Fly's hunt after a bear.

"When you and Brown took those two big bears out of the country and told of the eleven others you had seen, these fellows up here thought you didn't know a thing about hunting or you would have got the whole lot. What they know about hunting bear or anything else I don't know, but from the way they talked you would have thought the woods were full of bear-hunters. Fly, Beaty, Levette, and some others got up a big bear-hunt, and from the preparations they made one would have supposed they were going to have all the game in the woods. They were each going to have a bear-skin overcoat, and Rhea and me were each going to have a skin apiece for rugs. They borrowed every old shooting-iron on the place except that old cap-and-ball pistol of mine and started out. They wouldn't take provisions with them, for they said it was a mighty poor hunter that couldn't kill his own meat. They all 'lowed that the professor was a good deer-hunter, but he didn't know a little bit about bear. As they pulled out up the hill, Fly he hollered back to Rhea and said: 'When you see us coming down the mountain put the skillet on with plenty of grease, for we'll have bear-liver for supper.'

"Well, when they got away I took your old bear-trap down the cañon and set it. A bear had been around the night before where a calf had died, and I thought I might get him in the trap. We got a bear in the trap sure enough, and when they came down the mountain Rhea had a good mess of liver in the pan frying for them. Talk about hungry wolves. Those fellers were the hungriest set you ever saw. They hadn't had a square meal since they left, and as soon as they got in the house they began grabbing everything in sight. They couldn't wait for dinner, but took cheese, crackers, and everything else they could lay their hands on. Well, they sat down and began eating. They couldn't tell what kind of meat it was. They thought it tasted like liver, but knew I hadn't butchered, and when they asked, we told them it was the bear's liver they told us to cook."

The others here broke in with so many interruptions that Harvey was compelled to desist. But he had told his story, and now it was Fly's time to tell what he knew.

"I'll tell you that story straight. I've been listening to them talk for the last two years and I've got the whole thing straight. I'm onto the true inwardness of the whole business. We had a hard time up in the mountains after the bear we didn't get, and when we got back we had the bear-liver for supper all right enough, but you never saw two such scared people in all your born days as Harvey and his wife. I knew there was something back of their story, so I just kept my ears open, and this is the story of 'how we killed the bear.'

"Harvey set the trap down the cañon and an old foolish black bear came along and got his foot in it. Next morning Harvey went down to milk, and when he saw that the trap was gone he ran back and went clear through the house before he could stop. He dropped the milk-pails and went yelling at every jump, so badly scared that he almost fainted. Mrs. Harvey was so frightened that she couldn't say a word. Harvey ran to the gun-rack and then began cussin' us fellows for taking away all his guns. Then he tore around in the bedroom and out in the kitchen until Mrs. Harvey found her voice and asked: 'What's the trouble, High?' 'A bear! a bear!' was all the answer she got, for just then Harvey found his pistol and bolted off down after the bear. Mrs. Harvey followed with the axe, calling for High to come back or he would be killed.

"They rushed down to the corral where the trap had been, and sure enough it was gone. 'It was right here, Rhea. See, it is gone. Look where he tore up the brush. There he is—there he is! Get back, Rhea; you don't know anything about hunting bear. Get back, I say: if he should get here he would kill you,' Harvey kept yelling at the top of his voice, and then Mrs. Harvey would chime in, telling Harvey to keep back or he would be killed. She grabbed his arm and hung on, telling him not to go close to the bear, and then Allie came running down to see what it was all about, and Mrs. Harvey had a time of it trying to keep them both from the bear. Harvey shot all six loads from the revolver into the bear, but it never stopped him. Then they had to go back to the house to load up again. Mrs. Harvey begged High to let Allie go over to Hanson's ranch for help, and this made Harvey mad, because she thought he was not able to kill his own bear. Harvey put six more balls into the bear, and by this time the poor fellow got tangled up with the trap-chain and Harvey knocked him in the head with an axe."

"Well, I want to know!" said Harvey. "Do you believe that story, Dyche?"

"Well, I don't know how you killed it, Harvey, but I do know that you sent me a fine black bear-skin with a dozen bullet-holes in it. The skull had been mashed, too."

Dyche decided to make another trip to old Camp Bear Trail before the summer was over. The evening before the start was made Dyche took the shotgun, loaded with dust-shot, and went around the meadow looking for bird specimens. Eames was carrying the rifle and wandered off to one side. Suddenly Dyche heard the well-known "thump, thump" of a frightened deer as it dashed through the wood, and he hurried to the edge of the timber to get a look at it. An old doe was just disappearing in the woods, and about thirty yards behind her was a beautiful fawn. About the same distance back of the fawn was Eames, going at break-neck speed, shouting at the top of his voice, "Here, here it is, Professor!" The doe bounded over the pole fence, and the fawn ran along trying to find a hole to go through.

"Quick, give me the rifle," said Dyche.

"The rifle? why, I set it down by the fence and

tried to catch the fawn. I'd have got it, too, if it hadn't been for the fence."

"Catch the wind! Why, that fawn can outrun any man. The old doe would have been as easily caught."

"Don't you fool yourself. I'd have caught it, sure."

"Our old dog Jack has tried to catch that fawn a dozen times," said Allie, "but he never gets any nearer to it than you were."

A disagreeable incident marred the pleasure of the final visit to Camp Bear Trail, and Mrs. Dyche had occasion to learn by experience what a real mountain storm could do. While the train was passing through a tract of burned timber, where there was no shelter of any kind, a storm broke suddenly, the first warning being a clap of thunder. It rained a little and then hail poured down as if from an elevator chute. Eames and Dyche bunched the burros and covered them with rubber blankets, and under this hastily improvised shelter Mrs. Dyche crawled, staying until the storm was over. It lasted for half an hour and then continued with intermissions, alternately raining and hailing all the afternoon. It was late in the evening when a fire was started at the old camp, and, suffering with toothache, earache, and headache, Mrs. Dyche longed for civilisation.

Rain continued so steadily that they decided to return to Harvey's, and the party reached there, wet, bedraggled, and worn out. While making their arrangements to return to civilisation, the mountain streams were running torrents from the heavy rains. Over the top of Hermit's Peak rushed a magnificent waterfall, about eight feet broad at the top, changing to a fine mist before it reached the bottom of the cañon, a thousand feet below.

A group of Virginia deer.

CHAPTER VII.

In Indian Territory—A Frontier Fiddler—Life in a Dugout—Wild Turkeys, Wildcats, and other Wild Things—Surviving a Blizzard—An Indian Dance—An Embarrassed Professor—Successful Hunting—The Extinction of the Wild Turkey.

ON the approach of the next Christmas vacation, Dyche arranged to make a collection of the noblest game-bird in the world, the American wild turkey. With Professor Robinson, a veteran quail and rabbit hunter, he started for the Indian Territory, and a raw December day found the two at Caldwell, Kansas, wrestling with a mob of hackmen and omnibus drivers. This was the terminus of the railroad leading to the land of the Indians.

Engaging a light wagon, drawn by a stout pair of "buckskin" ponies, they reached Pond Creek just at dusk on the following day. A "dance" was in progress, and had been going on for the last twenty-four hours. A tall Arkansan, called "Short" on account of his size, was sawing away industriously at a fiddle, producing sounds which, by a good stretch of a vivid imagination, might be called music. The vigour of the dancers was evidence that his well-meant efforts were fully appreciated by the congregated cowboys and their partners. It was the event of the season, and visitors were there from ranches miles away. Cowboys and Cherokee half-breeds were out in full force, and, to supply a deficiency of ladies, dresses were put on several of the cowboys, who acted the feminine part to the satisfaction of all.

"Short don't play music outen er book," volunteered the lady of the house, "but he plays real tunes. He don't know a note from a cow-track, but he gets everythin' outen er fiddle there is in it, he does. He's ther best fiddler in the hull country, he is, and he allers stops till ther dance is done, too, he does. This is nothin' now. You jest orter bin here las' night. There were over fifty here, and ther cowboys thet danced for girls was real good ones, too. It was ther best dance we ever had. Some er ther boys got a little too much licker, but ever'thin' went off real nice."

Short continued his fiddling, and the cowboys kept time to the rhythm by patting their feet and hands, making such a noise that the tired travellers could not sleep. During the progress of the entertainment a deputy sheriff came in with a prisoner, whom he was taking to Caldwell to jail. The officer got so interested in the dancing that he forgot his prisoner, and the latter coolly walked out of the door and disappeared. He was not missed for some time, and then search was made for him with a lantern, but he made good use of his time and was not found.

Thompson's ranch, on the Cimarron River, sixty miles away, had been determined upon as headquarters for the hunt, and as the noise of the dancing prevented sleep, Dyche and the professor started early, and at daybreak were on the banks of the river. The road went directly to the water's edge, and could be seen emerging from the other side, almost immediately opposite. Confident that they were at the ford the hunters drove in. The water was soon up to the sides of the horses, and the wagon was almost floating. There was evidently a mistake somewhere, but there was nothing to be done but go on, and after a hard struggle the opposite bank was reached. They afterwards learned that the road went down the river some distance before it crossed, and then came back on the other side to the point where it could be seen.

Thompson's ranch was reached at dark. The house, or dugout, was a hole in a bank with a door in front, but no windows. It was filled with cowboys, who were very hospitable and helped to care for the ponies and got supper for the travellers. Pipes were brought out after supper, and the ill-ventilated room was soon so filled with smoke that Dyche and the professor, who were not smokers, were compelled to make frequent trips to the open air for a chance to breathe.

In the course of the evening it was learned that there was a turkey-roost about two miles down the creek, and one of the cowboys volunteering to guide the hunters, they made the trip. They had the pleasure of seeing a lot of squirrel nests, and returned to the hut after midnight.

Early next morning, with heads dull from sleeping in the atmosphere of tobacco smoke, Dyche and Robinson started for a hunt with cowboy guides. Robinson and Cimarron Jim went down the river, while **Dyche** and Buckskin **Joe took** the opposite direction. While working along through the patches of **scrub-oak** and over sand ridges, the latter two suddenly came upon a **flock of about** seventy-five turkeys. Joe at once put spurs **to his** horse and rode after them. Dyche was compelled to follow, and after a chase of about two hundred yards they were almost on the turkeys. They jumped from their horses, **but** before they could get their guns ready the turkeys disappeared over a hill. The race was repeated **with** the same success several times. At last Joe **shot** from his horse, and the flock flew and sailed out of **sight** over the thicket of oak bushes. A query from **Dyche** elicited the answer that this **was** one of the **chief ways** of hunting wild **turkeys** among the cowboys.

A mile further they came upon three deer feeding, and were **within seventy-five yards of them when** they were seen. Without waiting for the man whom he was engaged to guide, Joe jumped from his saddle and pumped bullets after the deer as **fast as he could** work the lever **of his** Winchester. The deer disappeared over the hill, but Joe affirmed that he had shot one through behind the shoulders, while he had **hit** another twice **as** it was going **over** the hill. **Tying** the horses to a swinging limb the trail **of the deer was** followed, but no sign of blood or a wounded **animal was to be** found.

The day was cold, but the unusual exercise of walking heated **Joe** to such an extent that he took off his overcoat and hung it to a tree. Half a mile further on he discarded his *chaperellos*, or leggings, and hung them up, intending to return for them on his horse. It soon became evident, however, that the cowboy was lost and could not find the horses. Dyche and Joe not being able to agree as to the proper direction to take, separated with the understanding that a signal shot was to be fired when the horses were found.

After an hour's walking Dyche found Joe's horse, loose and grazing, but his own was nowhere to be seen. His signal brought the cowboy, who immediately mounted the animal and rode off after his clothing, saying that he would return and take Dyche to camp. As night was now coming on Dyche did not wait, but started for the ranch, which he reached just at dark. It was three hours later when Joe came in, tired out and without either overcoat or leggings.

At midnight the noise made by some one stumbling into the dugout aroused everyone. It was Robinson, who was almost exhausted. He and Cimarron Jim had intended to stay all night at the hog ranch, but the guide got lost and wandered away, leaving the professor to look out for himself. Finding neither hogs, ranch, nor man, the professor made his way back to the home ranch, leaving his blankets with Jim. Knowing that the professor needed rest, Dyche gave up his sleeping-bag and said he would go out and hunt a little. As he shut the door he heard one of the cowboys remark:

"Well, that feller wants turkeys worse than I do."

The night was cold and frosty, and the stars gave sufficient light for Dyche to make his way up Turkey

Creek. He walked slowly, examining all the hawk nests and other bunches in the trees, which in the dim light took on odd shapes, and in some instances resembled turkeys. Everything was quiet, and not even the hoot of an owl broke the stillness. While thinking of the lonesomeness a scratching and scrambling on the ice of the creek attracted his attention, and Dyche saw a black object moving near a waterhole. Thinking it was a raccoon he sent a load of shot at it, and it spread out on the ice. Drawing it ashore with a long pole, Dyche tied its legs together and prepared to hang it to the limb of a tree, when the peculiar feeling of its tail induced him to light a match to see his prize. He found that he had killed a magnificent otter.

Wandering up the creek for an hour or two longer, he was just on the point of turning back when he espied a large object roosting in the top of a tall cottonwood tree. After watching it for some time he decided that it was a big gobbler and fired a shot at it, taking the best aim that he could in the dim light. The bird came down with a thump which told that the shot had been fatal. Tying its feet together and hanging it to a tree, Dyche found by the light of a match that he had killed a large golden eagle.

After dinner next day Dyche started alone for the jack-oak thickets. He had had enough of cowboy guides and severed all connection with Joe. He made his way slowly through the thicket, over the low sandhills, for two or three miles. Numerous places where turkeys had scratched away leaves searching for food were found. An occasional coyote was seen skulking through the bushes, and now and then he had glimpses of white-tailed deer. Golden eagles were sailing above, and quails and prairie-chickens were flushed in innumerable coveys. Finally a bunch of twenty-five turkeys was seen running over a hill, and a circuitous route was taken to head them off. A careful stalk placed him within forty yards of three of the birds which were scratching on a knoll, when the warning "pit, pit" told him he had been seen. A load of shot brought down a fine gobbler, weighing eighteen and a half pounds. As this was Dyche's first turkey and was a load of itself, the hunter was satisfied and immediately returned to the ranch.

Approaching the ranch, he heard a fusillade which sounded as if a battle was in progress. Thirty or forty shots were fired within fifteen minutes, and the men were evidently working their Winchesters as rapidly as possible. Cimarron Jim, who was cooking, had left his fire and was saddling his horse, while the others were not to be seen. Dyche and Jim were soon in the saddle and going in the direction of the sound. At the head of a ravine were the two other cowboys, riding and shooting. Jim rode straight for the scene, while Dyche circled around the head of the ravine, which he reached just in time to see a large wildcat emerge from the brush and start across the open space. Putting spurs to his horse, Dyche headed the animal off and started it in another direction, towards the timber. Again riding around it he drove it back, and while thus keeping it in the opening the cowboys on the other side of the ridge were almost splitting their throats yelling. Finally the cat made a straight run for the timber, and Dyche tried a flying shot at it with a charge of double-O shot, which tore up the ground all around the animal. The cat immmediately sat down and began to snarl and growl. Riding close, Dyche removed all the shot from a cartridge except about a dozen and killed the cat. One shot from the first load had hit it close to the spine, causing it suddenly to sit down.

The cowboys insisted that there was another cat in the thicket, and a search was made which soon drove the animal out. It came running along a cow-path towards Dyche, who waited until the animal was within twenty-five feet of him. Seeing the hunter the cat crouched in the cow-path. It was too close for a shot with the heavy load in the gun, and Dyche got down from his horse and began extracting the shot. The cat waited, showing its teeth and snarling, until a dozen number 3 shot ended its career.

Next day Dyche found a buck and doe and killed the former. Robinson came in with two fine turkey hens, and these, with several prairie-chickens and other small game, made a very respectable showing for the three-days' hunt. Robinson's time was up, however, and he was compelled to return to his home. Dyche accompanied him to Caldwell, and there prepared for a longer hunt. He made arrangements with three white men and an Indian for transportation to Fort Reno, starting the same afternoon on the journey of one hundred and fifteen miles.

Dyche had discovered that a man's blankets were considered public property, and in order to get rid of troublesome bedfellows and at the same time secure the greatest possible benefit from his bed, he had a sleeping-bag made of his bedding. The blankets were sewed together in the shape of a bag, and the whole was covered with heavy canvas. This "poke" excited the risibilities of his companions, but Dyche had the satisfaction of getting rid of certain insects which make very disagreeable sleeping companions. The first night's camp was made in the open prairie, and when the men saw how warmly Dyche slept while they shivered with cold all night, each vowed he would have a "sleeping-poke" as soon as possible.

About four o'clock in the afternoon of the third day a blizzard came upon them suddenly, while they were on the open prairie and far from shelter. The blinding storm soon prevented all travel, and they unhitched the horses, tying them on the leeward side of the wagons, while the boxes and bags from the load were piled up as a partial protection from the cutting wind. Putting their blankets together, the three white men lay down "spoon fashion" to keep warm. The Indian found an old buffalo-wallow and spread his blankets there in the high grass and weeds. Dyche followed the example of the Indian and crawled into his sleeping-bag, which he fastened down by the canvas. He was soon fast asleep, and did not awaken until he felt the Indian pulling at his bedding next morning, to see if he was alive.

The others were all alive, but were so cold and worn out that they could barely stir. Driving to the river a fire was started, and hot coffee and breakfast soon put new life into the party. Pushing on they reached Darlington, just across the river from Fort Reno, next day, Christmas eve. Dyche had a letter of introduction to the Indian agent, Colonel John D. Miles, who immediately took charge of the naturalist as his guest.

Colonel Miles was something of a sportsman himself, and readily gave information regarding the game of the Territory. His advice was that Dyche should go to Fort Cantonment, and arrangements were made that night with the driver of the buckboard that carried the mail, to take the hunter the sixty-five miles across the country to the fort.

The trip was to be made in one day, and the sour-visaged driver seemed doubtful as to Dyche's ability to stand it; but the latter said he could if any-one else could, and they started. After ten miles of the roughest riding Dyche had ever experienced, he got out a strap and fastened himself to the seat. The cold wind blew across the bleak prairie at the rate of forty miles an hour, and by the time the half-way dugout was reached the naturalist was chilled to the bone. The driver told him that they changed drivers and teams there, and that if he thought he could not stand the rest of the trip he could stay there until the next change, two days later. Dyche thanked him for the offer, but said he thought he could go the rest of the way. On a table in the dugout was a big corn-pone and an immense turkey, cooked to perfection. He needed no second invitation from the young man who was preparing to take the old driver's place, to help himself. With the drumstick and second joint in one hand and a huge piece of corn-bread in the other, he prepared to eat his Christmas dinner as he travelled.

The change of drivers was very acceptable, for the young fellow told many stories of interesting incidents of life among the Indians. While they were ascending a steep bank after crossing a small stream, the young fellow began peering into the bushes, and remarked that there were a good many deer in that vicinity. Suddenly he stopped and whispered:

"There's one now. Don't you see him?"

Looking in the direction pointed, Dyche saw a magnificent pair of antlers and a large body dimly showing in the bushes.

"Get down quick and get out your gun," whispered the driver. "He's a big fellow."

As the guns were strapped under the seat, Dyche answered that it was too cold and he was too stiff to shoot. Driving towards the deer the young man showed a cunningly contrived ruse. A deer's antlers were fastened to a stump and a gunny-sack formed the body.

"I have seen over a hundred shots fired at that deer," said he, and from the appearance of the trees and bushes in the vicinity he undoubtedly told the truth. The fort was reached just at dark, and Dyche was cared for by Decker, the Indian trader.

For three days a blizzard howled, and during that time all hunting was an impossibility; but Dyche spent the time of enforced idleness in getting acquainted with the Indians of the vicinity, and learning from them the condition of the country and the haunts of game. When the storm broke, Dyche took the advice of Little Raven, an Arrapahoe chief, and went to the big cattle ranch of Dickey brothers with Sam Horton, the foreman, who had been detained at the fort during the storm.

The ranch occupied a stretch of country fifty miles square, and the home ranch was a cluster of eight or ten log houses and stables, where the men congregated during the winter. This was headquarters, and was the base of supplies for the outlying ranches. Half a mile from this ranch was the winter camp of old Coho, a Cheyenne chief of great importance, and this camp was the place of resort and amusement for the thirty or forty men who made the home ranch their abiding-place during the winter.

The ranch was reached just before supper, and the meal was hardly over when the men began leaving by twos and threes, until the house was deserted by all except Dyche, Horton, and the old French cook, who, in response to Dyche's inquiries for the reason of this strange disappearance, said "the squaw humpers gone to the Injun dance."

Horton proposed that they should also go to see the fun, and the two were soon at the tepee of old Coho, which they entered without ceremony. Horton presented Dyche to the old chief, who sat on a roll of blankets between his two daughters, Zilpha and Cessonia. The chief was attired in buckskin leggings, with a blanket wrapped around his shoulders, while the two young squaws were dressed in pink calico gowns with red striped shawls thrown over their heads. Around the tepee sat other bucks and squaws, dressed much after the fashion of whites, with the exception

AN INDIAN DANCE.

that all wore moccasins and had blankets instead of overcoats.

After paying his respects to the old chieftain, Dyche followed the advice of Horton and purchased two pairs of moccasins from the girls, paying double price therefor, and thus winning the old fellow's favour. Following the presentation the whole party went to a large tepee, where the dance was to be given. In the centre of a room about thirty feet in diameter was built a fire in the most economical fashion, the sticks radiating from the blaze like spokes in a wheel. Around the wall, on rolls of blankets, sat about twenty-five squaws and two-thirds as many cowboys, with a number of young bucks. The sound of the "devil's fiddle," a peculiar drum made from a hollow log over which are stretched raw cowskins, was heard. Around this drum sat five Indians with short sticks, and they monotonously beat the drum in perfect unison, hitting it at intervals of about a second and a half in regular time, the "thump, thump" filling the whole room. To assist in the musical effort the five bucks set up a howl, prolonged, guttural, and undulating, rising and falling with regular rhythm and cadence. In this song the other bucks joined at intervals at their pleasure, while occasionally the squaws would unite their high falsetto voices in a most peculiar sound which they produced with lips and teeth and the tips of their fingers inserted in their mouths. The scene was weird in the extreme, and the darkened tepee, filled with a motley crowd of red and white men, sitting in the flitting lights and shadows of the fire and listening to the wild, barbaric music, brought back to the mind of the naturalist the scenes with which he had been so familiar in childhood.

The monotonous sound increased in volume, and then the signal for the dance was given. Two squaws selected one man, and the three took their places in the circle, continuing until the circle was complete. The ten cowboys and their twenty squaws then began a peculiar "crow-hopping" dance, varied with a heel-and-toe motion, each movement being made in unison, the circle rising and falling to the sound of the drum. For an hour the performance continued, and then an intermission was taken, while one of the musicians passed a hat—this was the only ceremony which he had retained from the teaching of the missionaries. The result of the collection being satisfactory, the performance began again, and continued with these hourly intermissions and hat-passings until daylight.

In the intermissions the squaws varied the entertainment by occasionally throwing their shawls over the heads of the cowboys, as an intimation that on the payment of a quarter the cowboy could have the privilege of kissing the squaw. To Dyche the price seemed exorbitant, but the cowboy taste appeared to be different, and they eagerly accepted the invitation. During the evening the naturalist had been an interested onlooker, with no idea of joining in the fun; but Zilpha and Cessonia had been so favourably impressed with his generosity in giving them double price for their moccasins that they asked him to be their partner in a dance. He was disposed to decline, but Horton advised him not to offend them, as they had paid a very high compliment to a stranger. Giving a reluctant consent he took his place in the ring, and the ludicrous figure cut by the hunter was such that Indians and cowboys kept up one continuous howl of laughter.

Horton, meanwhile, was scheming to have a little more fun at the expense of the innocent naturalist, and while the latter's attention was distracted, he gave Cessonia a dollar to catch Dyche and kiss him, while to Zilpha was given fifty cents to assist in the operation. Sitting by the side of Dyche, Cessonia suddenly whirled her shawl over his head and tried to draw him to her. He was too quick for her, however, and slipped to the ground and out of the shawl. Then began a race which afforded more amusement for the assembled cowboys and Indians than they had had for years. A second throw of the shawl, supplemented by Zilpha's strength, held Dyche. The two squaws, who had arms like prize-fighters, were more than a match for the naturalist, and they got him to the floor, where, after rolling over the ground from one side to the other, and almost tearing down the tepee in their struggles, Cessonia succeeded in planting a kiss all over one side of Dyche's face, from the mouth to the ear.

Fully satisfied with what he had seen and experienced, Dyche decided to return to the ranch, and to all of Horton's entreaties to wait for supper, which was just then brought in, he turned a deaf ear. This supper was a large kettle of meat and soup, flanked by stacks of thin loaves of bread, all of which was eagerly eaten by the assembled guests.

A wagon of supplies was sent to Loco camp, about fifteen miles away, and Dyche went there with a letter from Horton to the man in charge. The hunting outfit was shipped on the wagon, while Dyche followed on "old Weazel," a horse highly recommended for his good qualities. During the following ten days Dyche had no cause to regret his selection, for the horse was a perfect hunter.

He was now in the heart of the turkey country, and a preliminary skirmish that evening convinced him that there were several flocks in the vicinity. Early next morning he rode to Wolf Creek, four miles away, and while travelling carefully he suddenly came upon a flock of twenty-five or thirty turkeys, scratching under a bunch of jack-oak trees. He was seen, however, and in order to get around them he rode back out of sight, and then made a careful stalk. He was disappointed, and when he reached the place where the birds had been, they had disappeared. One turkey was soon seen running over a distant ridge, and Dyche hurried after it, getting another sight just as it was going over a second hill. A quick shot secured it, and with the sound half a dozen others rose from the grass and weeds, and one more was killed as they started to fly over the hill. Carrying his two specimens, which were fine old hens, to the horse, Dyche started for the ranch, satisfied with his first day's work. On the way down the creek he noticed another flock. Hiding in the bushes he watched them for over an hour, noticing especially their carriage and modes of living. He was satisfied with his study and a fine gobbler was killed. He took the three birds to the ranch, where they were prepared for future use. From the contents of the crops of the birds it was seen that the winter food consisted principally of acorns, a pint of which, shells and all, were found in each crop. With these were seeds of various plants, and one had eaten freely of wild grapes, which hung dried on the vines.

Next morning Dyche went again to Wolf Creek, and leaving Weazel feeding at the mouth of the stream in a grassy spot, he went up the creek on foot. A turkey calling in the distance attracted his attention, and while stalking it he came upon a large flock of over a hundred birds on a sand-bar, where they were sunning themselves. Some were scratching and dusting their feathers, while others walked about and picked up morsels of food. Occasionally an old hen would raise her head and give a loud call. Fully an hour was consumed in working around to a high bank, fifty yards from the flock, and here Dyche lay for some time watching the movements of the birds. Selecting a bunch of five, he sent a load of double-O shot into it. There was a roar as of a cyclone as that magnificent flock rose into the air and sailed away. He sent the load from the other barrel after them, and the double volley brought down four hens, which made a heavy load to carry back to the place where he had left the horse. On the way down the creek a fine gobbler was added to the bag, being shot on the wing as it started to fly.

Next day the same ground was worked again, and

two hens were killed out of a bunch which started out of the bushes. A few yards farther on a gray wolf came trotting down the ravine, and was knocked over with a load of shot; but before Dyche could get to him he recovered his feet and went into the brush. A large amount of blood was left, but the animal succeeded in getting away. Two more turkeys, fine gobblers, were shot on the return, after dark, from the trees where they were roosting.

The following day was lost by going out with the cowboys hunting deer, and the sixth day was so stormy that all were compelled to keep in-doors. Next day found Dyche again on Wolf Creek, and while watching for a flock of turkeys a wolf came down the ravine. A charge of double-O shot from "old Vesuvius" knocked him over, and this one did not get up again. A short distance further a flock of turkeys was scared out of the bushes and a hen and gobbler killed. These were hung to a tree and the hunt continued. Half a mile farther on a white-tailed deer was seen standing on its hind-feet, with its fore-feet swinging in the air as it nipped the buds from the lower limbs of a tree. The gun had turkey-shot shells in the barrels, and while the deer was not over fifty yards away, the hunter thought he would make sure of it by changing the shells. While so doing a shot slipped into the lock, but during five minutes of nervous work the deer did not get scared and continued to browse. Both barrels were fired to make sure of the shot, and the deer dropped dead within fifty yards of the spot where it had been eating.

Dyche now started back for the horse, and on the way down killed a gobbler. With the three turkeys, a deer, and a wolf on the horse he made his way towards the ranch; but before reaching it secured two more of the beautiful birds. As old Weazel had about as much of a load as he could well carry, Dyche walked the seven miles to the ranch, leading the horse.

Three more days in the woods added four turkeys and a wildcat to the total. The results of the hunt with his baggage were taken to Fort Cantonment in one of the ranch wagons, and just as he reached there a big freight wagon passed along from the apple ranch, going to Caldwell for supplies. The driver was lonesome and was glad of Dyche's company, so a bargain was soon struck and the luggage piled in the wagon.

Ten days of uncomfortable travelling, during which the whole outfit came near being lost by attempting to cross the Cimarron on the ice, brought them to their journey's end. While camped on the banks of the Cimarron, two cowboys undertook to have fun with the "tenderfoot," as they termed Dyche, and invited him to take a deer-hunt with them. Taking him to a bunch of jack-oaks they told him he would be sure to find deer in there, and advised him to crawl through and scare out the animals, while they would remain on the outside and shoot them as they ran out. Dyche noticed sundry smiles and winks, and fully understood the part he was to play in the little comedy. Crawling into the bushes he made his way carefully along one of the little trails.

While creeping in this way he saw a doe standing about thirty yards away, and a shot dropped her in her tracks. A buck, which had been lying down, jumped out, and a second shot brought him down. Tying his handkerchief to a bush to mark the spot, Dyche hurried out to where the cowboys were waiting. In a very excited manner he asked if they had seen the deer. None had passed that way, and when Dyche told them that he had seen two the excitement spread, and the cowboys now dashed into the bushes; but when they reached the dead deer with the handkerchief flying they concluded that the man they had sent into the bushes to play dog was no tenderfoot.

In this hunt after the finest game-bird in the world, Dyche not only secured a number of excellent specimens, but learned much about the habits of the wild turkey. He was aware of the fact that they begin to lay about the first of April, and laid from a dozen to fifteen eggs, but there was much about their winter life which he did not know and which this trip taught him. The birds of the Territory are slightly different from those of New Mexico, with brighter plumage and more distinct marking in colours. In summer the food of the birds is governed by the supply, being composed of insects, principally grasshoppers, and buds and berries. In winter the birds exist almost entirely on dried grapes, buds, and seeds, and especially acorns.

The cowboys said they had regular roosts, and showed Dyche places where they said they had seen thousands at a time in the trees; but the pot-hunters are after them, and this noble game-bird will go the way of the buffalo and be a thing of the past, unless vigourous means of protection are soon taken. The birds were formerly seen from Maine to Florida and from the Pacific to the Atlantic oceans, being more or less common along the wooded streams of the United States. Now they are scarce, and will soon be classed among the rarest birds.

CHAPTER VIII.

In the Cascades—Extermination of Many Species of Animals—Something about the Rocky Mountain Goat—An Arduous Journey—The Cascades Reached—Wholesale Hunters—In Camp—A Failure.

T is a fact well known to every reader that the American buffalo, or bison, is now practically extinct as a wild species. A few years more and this will also be true of such animals as the elk, moose, Rocky Mountain sheep and goats, caribou, musk ox, lion or puma, gray wolf and bears, while deer, foxes, and numerous other smaller animals will become very rare.

The moose formerly ranged from Maine to Idaho and was quite common in all suitable parts of the northern tier of States and Territories. During the past few years only a few stragglers have been reported in these places. It is but a few years since the beaver was said to be one of the most common wild animals in the State of New York. It is now questionable whether there is a single wild beaver in the State. Twenty years ago beaver were common along all the streams of Kansas. Ten years ago they were often found. Now they are rare, and in less than ten years there will not be a wild one, except by chance, in the State. Buffalo, antelope, deer, elk,

On guard.

bear, gray wolf, and the wild turkey have already disappeared.

With such facts as these before him, it was apparent to Dyche that if he ever expected to complete his collection of birds and animals it was high time to set about it. Following out this idea, he devoted a year to the study of taxidermy and to visiting the various large museums of the East as well as zoölogical gardens and shops of taxidermists. Some months were spent in the studio and workshop of W. T. Hornaday while he was mounting his large group of bison, and in the study and measurement of live animals in the various zoölogical gardens, and then he felt partially prepared for the labours before him.

In the *Century* magazine of December, 1884, appeared an article on the Rocky Mountain goat, from which the following extracts are taken:

"Its history is one of peculiar interest. So far as I know, specimens of the *Aplocerus montanus* are to be found only in three cities: In London, where an under-sized and wretchedly stuffed specimen does not redound to the honour of the British Museum or of English taxidermists; a better one in the Leyden Museum; two fair representatives, one male and one female, in the National Museum at Washington."

. . . .

"Of the twenty-three scientific authorities who have, so far as I have been able to follow the subject, written on this animal, none have ever seen one alive, and only four have ever examined a stuffed specimen, but they, nevertheless, have bestowed thirteen different generic names upon it, some making it a sheep, others classing it as a goat, while others again ranked it with the chamois."

So thoroughly was Dyche impressed with the necessity of prompt action in the matter of securing specimens that he did not wait for the board of regents of the university to consider his plan, but made immediate arrangements for an expedition to the far Northwest after big game. British Columbia was selected as the proper ground, and after a month's preparation he was ready to start. This preparation included daily target practice with two new rifles: a Winchester, 40-82 model of 1886, and a 40-75 Remington. Day after day the practice was continued, until over forty pounds of lead had been shot at every conceivable form of target, under every possible circumstance which his experience had taught him he might expect in the woods of the Cascade range. Rolling and stationary targets were fired at from a standstill or while running; he would run at full speed forty or fifty yards and then fire, until he became familiar with his guns under all circumstances and until he felt that he had full control over his nerves and muscles as well as a perfect knowledge of just what his guns would do. Tarpaulins, for covering packs and drying skins, heavy pack-bags and smaller sacks of canvas were made, every possible want was cared for, and on the morning of July 3d he left Lawrence, reaching Denver on the 4th.

A few days were spent here waiting for Judge Frederick A. Williams and Dr. J. W. Anderson, who were to accompany him with a local taxidermist who professed to be able to guide the party through the Cascade Mountains or **anywhere else**. During this wait in Denver Dyche continued his target practice, and on the eleventh of the month all were ready and the start made for the north.

The hot, dry weather made the long trip through Colorado and Idaho anything but pleasant. At Montpelier the train stopped half an hour for breakfast, and was just too late to allow the passengers to witness a bear-hunt in the heart of the **town**. A large female grizzly, with a cub, had come to town and walked leisurely through the streets. The whole population turned out, and for her boldness and contempt the bear lost her life, while the cub was made prisoner, and when the train arrived was showing its viciousness and strength by whipping the best dog in town.

The train sped on over the sage-brush country. Indians were to be seen standing at the stations watching the iron horse which had invaded their domain. Finally the region of sage-brush disappeared, and as Pendleton was approached the whole country took on the appearance of a vast wheat-field. Trees, water, houses, and barns were seen, and civilisation seemed to have been reached once more.

Spokane was the end of their railroad journey, and here tents were put up, bath-houses and barber-shops invaded, and a day of rest on Sunday made the travellers feel once more civilised. Monday and Tuesday were spent buying eight or ten horses. After withstanding the wiles of the dealers, who apparently made their living by cheating "tenderfeet," a good outfit was purchased and the horses and wagons were made ready.

Learning at Spokane that a good wagon road led to Palmer Lake, it was determined to go on at once, and the doctor and the judge went ahead, leaving Dyche and Mac, the guide, about a day behind. The road lay through a country where there was but little grass and water, and it was ten o'clock the first night when Dyche and his companion reached a ranch. The task of keeping off the horde of vicious dogs and waking the ranchman fell to Dyche, who accomplished it after much trouble. A little wheat-straw was obtained for the horses, and next morning, when Dyche went to the house to apologise for the trouble he had given them, he was invited to a fine breakfast.

All day the road led through a hot and dry country. Dust and sand from five to ten inches deep made the way almost impassable. When the wind blew it felt as if it came from the mouth of a furnace and the hot dust flew in clouds, completely enveloping the wagon and pack-horses. The whole country was as dry as a desert, covered with large stones, and there was little or no vegetation to be seen anywhere. Late in the afternoon a ranchman sold them enough wheat-straw to make a partial feed at three cents a pound, but he would let them have no water at any price. His well was nearly dry, and he gave them a little to drink, but the horses had to go without. Camp was made, but at three o'clock next morning they started and travelled to Wilbur, a small town, where, at six o'clock, they got sufficient water for the horses and then started on, overtaking the doctor and the judge at eight o'clock.

All day long they toiled on through the sand and dust, which was now supplemented by clouds of smoke from the distant burning mountain. Camp was made at noon and an attempt made to rest and eat, but it was a dismal failure. A little water found in a hole, which was so vile that the horses would not drink it, was boiled and coffee made, but it was not fit to drink.

At half-past three in the afternoon the Grand Coulee was reached. Here the road dropped down almost perpendicularly into the cañon and rose as precipitously on the other side. In the distance could be seen the magnificent range of the Cascades, along the Columbia River, which at this place flows through an immense gorge. Here the party suffered the tortures of Tantalus, for while they could see and hear the water as the stream dashed down the gorge far below them, they could not by any possibility get down to it, and were compelled to travel three miles farther before they could obtain a drink.

After a short rest the long pull up the mountains began. The saddle-horses were fastened by a long rope to the tongue of the wagon, and thus assisted in getting the vehicle up the trail. The road narrowed and wound in and out among the steep hills, along the crags and cliffs of the mountain-side. While going over this trail, Mac's bucking "cayuse" took it into his head to show off, and the result was that he fell and got tangled up in the ropes, dragging down Dyche's horse. The latter was almost caught in the fall, but extricated himself. Then the doctor's horse got down, fastening his rider by winding the rope about his legs so that escape was impossible. The struggling and kicking horses were on the edge of a precipice which would give them a sheer fall of three hundred feet should they go over, and the situation was precarious in the extreme. The ropes were quickly cut and the doctor relieved from his dangerous position, and then the horses got up safely.

But these were not all the troubles of the trip over the pass. A few miles farther on a freighter stuck fast in the mud effectually barred the way. It was utterly impossible to get past the big wagon, so three hours were spent in assisting the freighter and clearing the road. Camp was made that night by a small spring, and by six o'clock next morning they were again on the road. At noon they reached Condon's Ferry, the home of "Wild-Goose Bill," who spent several hours, when he first came into the country, stalking and killing his neighbours' tame geese.

The early morning breeze increased to a hurricane and the dust filled the air, while heavy black clouds of smoke and ashes covered the sky. The discomfort of travelling was so great that camp was made early in the afternoon, and they undertook the almost impossible task of getting supper amid the dust and ashes. Dyche and the judge started a fire and cooked a supper after a fashion, but everything was so mixed with ashes that eating was almost impossible. The doctor was the life of the party, and his exuberance kept them from utter demoralisation.

The potatoes were half fried and full of dirt. "All

the better for that," was the doctor's comment, and he ate as if he believed it. The bread was burnt. "All the better for that," was the invariable remark, and at last the expression became the general byword whenever anything went wrong on the trip.

The clouds passed away during the night and the winds died down without a drop of rain. The whole country appeared to have had no rain since the flood. It had rained a little, however, about four months before this date. All next day was spent in one continuous trip over roads in valleys between the mountains, while the dust came in clouds and covered everything. With goggles over the eyes and with the mouth and nose covered to keep the fine particles from the lungs and throat, they almost suffocated as they travelled, until late in the afternoon, when they reached a small lake, where Dyche succeeded in shooting six half-grown ducks. While thus engaged an Indian stole his spurs which he had left on his saddle, and as he could not spare the time to chase the thief, his horse benefited by the episode.

Reaching the Okonagan River at a point where it is about one hundred and fifty yards wide, they found that a band of Indians had made their camp and were fishing for salmon, which were running in great numbers. The red men erected a fish-dam of willow and pole in such a manner that all the fish going down the stream were caught in the baskets. This method has been in use among the Indians from time immemorial and was fully described by Lewis and Clarke in the history of their explorations in 1805.

Between two and three hundred **Indians were in** the camp and they secured from one to two thousand pounds of fish each night. These were cut up and placed on pole platforms, under which smudge fires were kept going constantly by the women and children. The combination of squaw, papoose, dog, fish, and general Indian camp smell was something better imagined than described, and little time was wasted by the hunters in that vicinity.

Ward's ranch was reached late that night, and it was nine o'clock before the ducks were dressed and cooked. They had a strong fishy taste, but all adverse comment was cut off by the doctor's "They are all the better for that." The judge, however, rose to the occasion, saying:

"Begorra, I like fish, and, begorra, I like ducks, but, begorra, I do not like fish-duck."

Leaving Ward's ranch at half-past seven next morning, another hot and dusty day was experienced. White Mountain, that landmark known to every man who has travelled in the Cascade range, towered on the right of the trail and stood out against the smoky sky. It was an inspiring sight and aided in whiling away many of the tedious minutes of the day. The party had been told of the lake called Sweetwater, and it was with a feeling of joy that the waters of the place were seen, but how keen was the disappointment when it was found that some facetious prospector had given the name to the bitterest sheet of water in the Northwest. They toiled on and just at dark reached Loomis trading post, utterly fagged out.

This post was composed of a number of log-cabins on the banks of a beautiful ice-cold stream which dashed down from the snows of the Cascade range, to which the early French voyageurs had given the name "L'eau de coulée," or "water of the chasm." The trappers and prospectors who followed at a later day, however, corrupted this into "Toad Coulee," by which name the stream is now known. The post was patronised to the extent of fifty cents for a fine supper, prepared by a good cook, and the horses were regaled with alfalfa hay in the corral.

During the moments of rest in their long and wearisome day, Mac had exercised his ingenuity in getting up new stories about the beautiful fish which were to be caught in Palmer Lake. Now that the body of water was so near, the doctor and the judge wanted to hurry on and try a cast. Within a mile of the lake the wagon stuck fast in the mud, and they were compelled to pack the camp equipment to the shore. Once there, the two fishermen could not even wait for lunch, but soon had their lines out and were whipping the water in vain effort to bring a rise. In about two hours they succeeded in getting a dozen of the worst specimens of fish to be found in the whole country. They were little things called "white fish" in that country—and were so full of bones as to be unfit for eating. There was not a trout in the lake.

Evening was now coming on, and the horses, which had become frenzied from the attacks of the flies and mosquitoes, were taken high up on the mountain, where they could have relief during the night. It was useless to try to sleep, for the winged pests made life a burden the whole night long. At earliest daylight the party was up, and scarcely waiting for breakfast, they struck camp and went back to Thorp's ranch, where arrangements were made to store the bulk of the outfit.

After an excellent dinner served by Thorp's squaw, they started once more for the hunting-grounds. The trail now led up the side of old Mount Chapaca, and was a most pleasant change from what had been experienced during the past few days. The air was clear and cool, and there was none of the dust and heat which had made life a burden during their long ride over the barren country. Just as the sun sank behind the western range a cosy grassy spot beside flowing water was seen, and as the horses were halted for camp it was discovered that the place had been utilised for the same purpose by someone some time before. On a large pine tree was found the legend: "Camp Disappointment. Could not find the hidden mine."

A fawn had fallen a victim to the rifle of one of the party as they came up the mountain, and by the side of the cool spring the camp was anything but a disappointment to the naturalist and his companions. Mac had been telling of a certain beautiful stream just ahead which was full of trout, and next day the place was reached early in the forenoon. Mac's fish-stories were generally looked upon with suspicion by this time, but the doctor and the judge were so inspired with his apparent truthfulness about this river that they did not wait to eat, but hurried away and began casting their flies **over the stream. After** whipping the water for an hour or more they gave it up in disgust and returned empty-handed to camp. It was afterwards learned that Mac had not broken **his record,** for there was not a trout in the stream, and furthermore there never had been, for a series of high falls some distance below prevented the fish from coming up stream.

Several deer were seen during the afternoon, **and** just as evening **was** coming on the smoke of a campfire was espied, which proved to come from the camp of two **deer-hunters.** The **two** men who **sat by the** fire **eyed the** new-comers with suspicion, **and** in response to inquiries said they were prospectors; but the piles of deer and sheep skins which lay about the camp made it evident that they were skin-hunters, and were in the mountains ruthlessly slaughtering the game for the paltry sum to be had for the hide. It was afterwards learned that these two men killed over two hundred deer during that summer, selling the skins for seventy-five cents each and leaving **the** dead carcasses lying just where they had fallen. The head-hunters, against whom such a wave of indignation has spread over the country, are bad enough, but they are usually satisfied with two or three animals, while the skin-hunters indiscriminately slaughter the animals by the hundreds and soon drive them **out of** the country.

The skin-hunters **were** full **of** advice when they learned that the party was after sheep.

"You may see some of them on the highest peaks **if you** keep your eyes skinned," **said** the man with a buckskin suit, "but you can depend on it, Mister, that one old ram will always see you. He's jest sure to see you if you're in sight. One old ram is always on the lookout on the top crag, while the others of the band are feeding below on the slope."

"How did you get two of them if they are so shy?"

"Oh, we came upon them while they were in the woods and didn't expect us. Then we fellers have had lots of experience hunting wild things, and can get animals you fellers can't."

This big talk did not have the effect of deterring the party, for Dyche believed that if anyone else could get a sheep he could. A mile's walk took the naturalist and his companions to the ideal spot for a camp, and here preparations for a long stay were made. A mile to the north two towering peaks raised their heads to the sky, while between them a clear, ice-cold stream dashed down over the rocks and bowlders, making constant music through the grassy valley. On the west side of the stream the valley sloped gently back to the higher mountains, while the thick grass made a carpet of green. Midway between the water and the rising ground stood a bunch of spruce trees, forming a background for the camp, which was pitched just to the east of the trees. The horses were put on ropes, the tents set and carpeted with spruce boughs, the fire started and supper begun, and the naturalist and his friends were at home.

The judge confessed that he did not know much about cooking, and had been afraid to try his hand heretofore lest he should spoil the meal; but he had such an appetite after the long trip that he could not wait and began preparing to cook. He got out a lot of flour, and notwithstanding the fact that he received enough advice to sour any batch of bread, he succeeded in making such biscuits that he was at once appointed baker for the expedition.

The doctor not only knew how good coffee tasted, but knew how to make it; and his efforts in this line supplementing those of the judge, it was but a short time before the party sat down to a feast so fine that all else but the luxury of dining was forgotten. Everyone was hungry, and the table was soon cleared.

As they were in a sheep country, the high crags and peaks pointed out by the skin-hunters were eagerly scanned in the hope of seeing the longed-for game. Early next morning the first hunt was taken, but it resulted in nothing but experience; and that evening Dyche showed such signs of breaking down under the unwonted journey and horseback travel that the doctor insisted that he should remain in camp for at least two days. Each day the judge and the doctor would go out on the mountains and return in the evening with no game, but with such stories of the sheep which they had seen that Dyche fretted at his enforced idleness. On the evening of the second day the doctor came in with a story of a band of sheep which he had seen, and at dawn next morning he and Dyche were off to the spot. By sunrise they were on the top of a magnificent mountain, and saw the orb over the peaks of the range.

Separating from his companion, Dyche wandered

along slowly, watching far in front and looking for any possible movement that would indicate a sheep. Each rocky crag as it came into view was carefully scanned with the expectation of seeing a ram on the alert. In a grassy cove on the side of the mountain unmistakable signs of sheep were seen. He passed on, looking carefully in every direction, keeping near the top of the ridge. Finally he retraced his steps to the cove and made another inspection. It was one of the places which had been pointed out by the skin-hunters as "a likely place for a few old rams," and the hunter was loth to leave it. While moving along near the crest of the ridge, Dyche heard the sound of a stone rolling down the mountain on the opposite side of the hill.

The sound could only be made by some animal walking among the loose rocks, and in an instant the hunter's heart was leaping madly at the thought of his nearness to the "big-horn." The wind was in his favour, and as the sound continued he felt sure that his presence had not been noted by the animal. Removing his shoes, he crawled towards the top of the ridge as carefully as a cat crawls upon its prey. The movements of the animals became plainer with each foot that he advanced. Just as he reached the crest, he stopped for breath and looked to see that his gun was all right. Again the animals moved. They now seemed to be only a few feet away. Slowly he crawled a few feet farther and rested. Now he began a snake-like movement at almost snail's pace towards the crest. A few feet more and yet they were not in sight. Another move and he was behind a jutting rock which stood on the top of the mountain. The animals were still moving about almost within reach of his arm, yet he could not see them, although they were heard so plainly. He lifted his head carefully, just enough to permit him to peer through a crack in the rocks, and there, standing in the sunshine and not forty feet away, almost under him he saw, broadside towards him, two big mule deer bucks. The revulsion of feeling at this moment was so bitter that in his disappointment he jumped to his feet and began throwing rocks at the deer, which went down the mountain faster than they had ever travelled before. Dyche was so sure of finding sheep when he lifted his head above the rocks that his keen disappointment unnerved him, and he returned to camp disgusted.

CHAPTER IX.

The First Big-Horn—How his Skin and Bones were Prepared—Habits of the Rocky Mountain Sheep.

YCHE'S disappointment did not prevent an early start next morning, and it was still early when he reached the spot where the bucks had deceived him. After climbing over the mountains for some time he reached a semicircular ridge, opening to the south, through the centre of which ran an intersecting ridge, terminating in a jutting crag of bare rocks. The top of this ridge was bare and rocky, while short and thick bunch-grass grew on the sides of the slopes leading down to the central amphitheatre, towards the woods. From the dense growth of timber to the south of this horseshoe-shaped inclosure there was a gravelly hill leading towards the central ridge which divided the amphitheatre into two parks. The three jutting spurs at the ends of the side and central crags were a quarter of a mile apart, while from the central crag to the back of the ridge it was fully half a mile.

Dyche had worked his way up the mountain and it was noon when he reached the top of the ridge. Crawling carefully to the top, he examined the country thoroughly before showing himself. As nothing living was visible, he crossed the ridge and sat down on the south side, overlooking the central portion, and began eating his lunch. While thus fully exposed to anything coming from the south, he saw a big ram walking from the woods up the gravelly hill to the crag at the end of the central ridge. Dyche knew that if he so much as moved his hand or foot he would be seen by the keen-eyed animal. He thought rapidly now, for it was a case of no sheep if he moved and perhaps no sheep if he did not move. The ram reached the crag, and after gazing at the country for a while began feeding towards the spot where the naturalist lay.

Dyche saw that his only possible chance would be to sit perfectly still until the old fellow got close enough and then shoot him. The ram was wild and had evidently been hunted before. He would not take time to graze, but would snatch a mouthful of grass and then raise his head high in the air and look about while he chewed it. He continued slowly towards Dyche, but the naturalist's clothing was of a colour that was indistinguishable from the rock and earth on which he lay, and he was not seen. The ram fed towards him some fifty yards, and just as he was beginning to congratulate himself on the success of his plan the animal suddenly pawed the earth a little and lay down, facing towards the naturalist. There he contentedly chewed his cud, while Dyche hardly dared to breathe for fear he might be seen.

Minute after minute passed and the sun slowly crawled towards the western horizon. At last the sheep got up and shook himself, and Dyche felt that he would now surely come on along the ridge, but in this he was again disappointed, for the sheep began to feed again, but edged around the base of the crag towards the woods from which he had come. The naturalist saw that if he was to secure that sheep he must do something very soon, or it would be too dark to see to shoot. Timing the ram, Dyche would give himself a shove with his heels every time the sheep's head went down after a mouthful of grass. Then drawing his gun up he would wait for another mouthful and give himself another shove. In this manner he covered the few feet between himself and the top of the ridge in half an hour and gradually worked himself over. As soon as he was confident that he was out of sight of the animal, he made his way along the eastern spur of the ridge to the southern end. Making his way carefully to the top, he peered over and saw that the ram was still there, but was fully a quarter of a mile away.

Retracing his steps, Dyche made his way clear around to the south end of the western ridge, but he was still as far from his game as ever, and he could see no way of getting closer without exposing his body. He now went to the spot where he had first gone over the ridge, and waited, hoping the ram would come towards him, but he soon saw that the animal was going towards the timber. The sun was now almost down, and the naturalist saw that what was to be done must be done quickly, and concluded to make a desperate effort to get that sheep. Divesting himself of his shoes, hat, and hatchet, he hurried back along the western ridge to the southern spur.

When the ram would reach down for a mouthful of grass Dyche would crawl along, as rapidly as possible, down the side of the ridge into the little park between the two ridges. He took advantage of every rock and hole in the ground, and in this way reached the foot of the central ridge in about fifteen minutes, while the ram was feeding just over the crest and out of sight.

Now came a series of movements between gliding and crawling towards the top. Dyche had marked the spot where the ram was feeding and worked his way to it, getting closer and closer to the place where the ram had last been seen. Just as he was working to the top of the ridge he heard a scrambling noise on the other side. Springing to his feet, with his gun forward and ready, he saw the ram going at full speed towards the timber. A hasty shot and the animal was seen to falter, but quickly gathering itself together it went on. Another quick shot and the ram disappeared around the edge of the crag. Dyche ran to the end of the ridge, where he had a clear view of the slope leading to the woods, but not a sign of the ram was to be seen. A search along the edge of the ridge showed the old fellow standing about seventy-five yards below, apparently hard hit. Dyche's gun was at his shoulder almost instinctively, but the shot was not fired, for the sheep gave a lurch forwards and went tumbling down the side of the mountain. Hurrying after him, Dyche found a magnificent specimen and hastened to take advantage of the fading light to make anatomical notes and measurements. He found that the eyes, which the books describe as being brown or dark hazel, were of a straw colour with a slight mottling of hazel near the edges.

Darkness put a stop to the examination. There was neither wood nor water near or Dyche would undoubtedly have camped near his first big-horn. An hour was spent in finding his shoes and hat, and eight o'clock was past when the naturalist reached camp. A hot supper was waiting for him, and while discussing it the hunter told of his adventure.

Everyone was up bright and early next morning. The doctor went south, while Dyche saddled Billy and went after his sheep. Two hours were spent in measuring and skinning the animal, and by one o'clock the skin, skeleton, and most of the meat were in camp.

The doctor arrived from an unsuccessful hunt in time for dinner. The sheep steaks proved that the flesh of the big-horn is the finest game meat in existence. Dyche lived for weeks on the flesh without having it pall on him, which shows that it is different from any other venison. The meat was tender and juicy, having only a slight mutton flavour, while the fat, or tallow, would not harden, but formed a granular mass, except in the coldest weather.

The afternoon was spent in preparing the skin for preservation, which was a simple operation. All particles of flesh and fat were first thoroughly removed from every part of the skin to the hoofs, and then a thin-bladed knife slipped between the hoof and bone. The ears and nose were cleaned of fascia and cartilage. Four parts of salt and one of alum were placed in water, and the whole was boiled until a strong brine was made, which was allowed to cool. An excavation was made in the ground eighteen inches in diameter and six inches deep, and into this a part of the skin free from bullet-holes was pushed, forming a cup, into which was poured the milk-warm brine. Then the whole skin was thoroughly wetted with a small sponge. The head and feet were then placed in the vat and the whole skin thoroughly saturated with the brine and left to soak for six hours, when the operation was repeated. When this soaking was finished the skin was hung in a shady place to dry, care being taken to turn the edges out frequently. At the end of a few days the skin was dry and ready for packing. It was folded and sewed in a burlap bag marked with a label showing a number which corresponded with the number of the description in the note-book. Each bone was also marked with a similar number and the specimen was ready for the storeroom. This same process was gone through with in every case where the skin was preserved.

Just as the sun was going down a big animal was heard in the woods tearing along at full speed, making a noise like a herd of deer. The doctor and Dyche jumped for their guns, and were ready for anything, when out of the timber came the judge on Nellie Gray, at full speed, waving his hat and shouting.

"Well, by the long-horned angora!" said the doctor, "I'll bet the judge has killed a deer. Yes, there it is hanging to the back of his saddle."

Nellie Gray seemed to share the excitement of the judge as they came tearing into the camp.

"What did I tell you?" shouted the judge as he jumped to the ground. "Just look at that specimen, will you? Whoopee! A hundred dollars would not buy that sheep."

As he spoke he began unloading the head and skin of a fine ram from his saddle. While the naturalist and the doctor prepared supper the judge told his story.

"Do you see that mountain off there? It's more than five miles over there and the trees are down dreadfully between here and that old bald-top. I rode all the way, though, and Nellie had a terrible time getting over the logs. At ten o'clock I tied Nellie at the foot of the mountain and climbed to the top. I went along that snow-bank which you can see over there, and as I slipped along I saw a sheep. I worked around a crag and got within seventy-five yards of him. There was a whole band of them and they did not see me. That head and skin tells the rest of the story."

Williams was so elated over the fulfilment of his prophecy of the night before that he at once proceeded to perform the ceremony of christening the camp. In honour of the queen, in whose dominions the party were encamped, it was called Camp Victoria.

After several days' unsuccessful hunting Dyche concluded to extend his range and went to a bald ridge about six miles from camp. Ascending the highest peak in the whole country, from which a vast expanse of mountain and valley could be seen, he carefully scanned the surroundings. He espied a moving object about a mile away and finally made it out to be a sheep. It was walking along the side of the mountain, and the naturalist dropped to the ground and made a long stalk to the place where he thought the animal would pass. He crawled to the top of the ridge and waited half an hour before he saw the sheep, between two hundred and three hundred yards to his left and about one hundred from the top of the ridge. Near this sheep could be seen two others, lying down. He made a careful stalk to the ridge just above them, but could not see them. After waiting in vain for the animals to move along the mountain he rolled a stone down in order to start them out, hoping to get a flying shot at them as they ran. Either the sheep were accustomed to rolling stones or their hearing was not as good as their sight, for they did not move.

A larger stone was then started down, and this did the business effectually, for it started others and the whole mass went bounding down, over the ledge, right among the sheep. The animals went at full speed along the side of the mountain to escape from the rocks, and as they ran Dyche got in two shots. He now ran to another point where he could see the sheep rounding a crag. There were only two running now, and they got out of sight too quickly to give him another chance. Going back along the side of the mountain, he saw a beautiful four-year-old ram just as it was sinking to the earth. By the time the measurements were taken and the animal skinned it was after sunset, but the load was carried two miles to the horse. Dyche was greatly exhausted from the exertion and thirst, but was compelled to walk another mile before he could ride. It was late at night when camp was reached, the horse taking his own way.

The doctor and the judge had been "hunting large," seeing many sheep, but the latter were so shy that they could not get them. They each succeeded in securing a fine specimen, and now their time was up and they had to return to Denver.

Dyche had been dreading this announcement for some time. He felt that he could ill afford to lose his friends, for two better camping companions could not be found. Though both were professional men, with large business to look after, they were thoroughly versed in woodcraft and were real campers and hunters. Having a lively interest in everything going on, they took every disappointment the weather or country might bring, without complaint. Evening in camp was the most pleasant part of the twenty-four hours, for it was spent in profitable conversation, exchanging ideas and constantly developing new thoughts.

A general discussion on sheep was started the evening before the Denver gentlemen left for home, when the judge asked if it was necessary to travel away up into the British-Columbian mountains to get sheep.

"No," said Dyche. "Sheep range from New Mexico to British Columbia, and may even be found as far north as Alaska. They are seen east as far as the Black Hills and range west to the Pacific Ocean. They live, however, only in such localities in the prescribed territory as best suit their habits. At present they are confined to a few favoured localities in the highest and roughest parts of the Rockies.

But I learn that the skin and head hunters are fast thinning them out. It is only possible now to find stragglers, and these will soon be gone."

"How is it that your sheep are so much darker than mine?" asked the judge.

"The colour is difficult of description, as it varies so much in individual specimens. Some are very light grayish-brown, or light rufous gray, while others are very dark. All the intermediate shades from light rufous ash to dark chestnut are to be seen in one band. There is always a dingy white patch on the rump, like that of the elk or antelope."

"Well, there is one thing that I can't quite understand, and that is why we don't see any ewes or lambs," said the doctor. "Here we have been hunting for several days and have seen nothing but bands of old rams. It looks as if all the females had left the country or stayed in the thick woods."

"During the summer and autumn the rams range together on certain mountains, while the ewes and lambs are in separate bands on some other range, where they stay until late in the fall. You will find that there are plenty of ewes and lambs within fifteen or twenty miles of these mountains. Not over half of the ewes have lambs following them. I have tried to find the reason for this, but have not been able to satisfactorily account for it. Among the theories of the old hunters the most tenable is that the lambs fall victims to their natural enemies, such as eagles, wolverines, and wolves."

"Do the sheep remain constantly above timberline?"

"They seem to like the high crags and mountain-tops, and when undisturbed remain there most of the time. They grow very fat on the short, thick bunch-grass that grows on the slopes and coves on the mountain-side. The high ground affords them the opportunity to watch for their enemies. When disturbed they always break down the side of the mountain for the woods, but usually keep going until they reach another range of high mountains, though it may be miles away.

"Ewes and lambs do not range on the high mountains, at least while the lambs are small, but remain lower down near the edge of timber-line. Certain alkali spots on the side of the mountains are great places of resort for the sheep, and they go there as frequently as deer do to a salt-lick. The lambs are born in the latter part of May or the first of June."

"There is one thing that always seemed a myth to me," said the judge, "and that is the stories we hear about the fearful leaps rams make down precipices, where they are said to alight on their horns and rebound to their feet, thus saving their legs from the terrible shock. I never believed there was any truth in such stories."

"Well, as usual, your judgment is correct. Those stories are all myths evolved from the fertile brains of those men who do their hunting by the fireside of some ranch in the mountains. Take a man who comes to a place like Thorp's ranch and shows the people that he is a 'tenderfoot' and is going to write a book, and they will fill him up with more stories of adventure than a hunter can find in a lifetime.

If you will watch an old ram going down a mountain, you will see that he does not jump down steep places, but is as careful as a dog about where he puts his feet. He will feel his way down and slip and slide, keeping a firm foot-hold all the time, and never jumps any more than any other animal which ranges the mountains. I suppose that the imagination of some book hunter made the fearful leaps out of the battered condition of the horns. He possibly could not understand why nature wanted to put such horns on an animal, and not knowing that the horns had been battered up by fighting, he imagined that it was done when the animal jumped and struck on its head.

"The horns of the males are of immense size, but the ewes and lambs have small ones. Now, the females and lambs have to jump and go where the rams do, and if they jumped and struck on their horns they would have a sorry time of it. The horns of the males vary from twelve to eighteen inches at the base, and a cross-section shows that they are all triangular in shape. The horns and skull of the largest sheep I ever saw weighed, when thoroughly dry, twenty-eight pounds. The horns of the largest males average from thirty to forty inches in length, while those of the ewes are rarely over twelve inches long."

Early next morning the judge and the doctor started over the trail to Thorp's ranch, leaving Dyche practically alone in the mountains, for the guide was no companion and took little interest in anything beyond his immediate wants. The naturalist now made great progress in his work, for he went at everything with all his might in order to drive all thought of lonesomeness from his mind. He collected many small mammals and birds while hunting for the larger ones.

CHAPTER X.

End of Sheep-Hunting—How the Sentinel Fell at his Post—
A Peculiar Wound—Finding the Noon Hour by Stars—
How the Collection of Sheep was Completed.

THUS far all the hunting had been done within five miles of the camp, but now the naturalist determined to take a wider range. Under ordinary circumstances a mile or two is not considered a great distance, but when it is over the roughest range of mountains in the world, it generally means a day's constant struggle to get over the ground without devoting any time to hunting.

At daylight Dyche was on old Charlie's back, and ten o'clock found him eight miles north of Camp Victoria. The horse was lariated out where he could feed, while the hunter climbed to the top of the highest mountain in the vicinity. From this eminence the whole country was carefully scanned, with the expectation of getting a glimpse of a sheep. At last one was discovered about a mile away on the side of the mountain. The colour of these animals so nearly approaches that of the rocks and dirt among which they feed that it is almost impossible to distinguish them unless some movement reveals their whereabouts. Careful scrutiny of the place where the moving object was seen developed the fact that there was a band of the animals, but as they were on a barren mountain-top, four hours were vainly spent in the endeavour to get close to them without being seen. Finding this to be impossible, Dyche returned to camp, as he preferred to leave them for another day when they might possibly be in some more accessible place.

At nine o'clock next morning he was back again, carefully scrutinising the rocks and hills, and at last made out what he thought was the head of an old ram on the top of one of the highest crags in the vicinity.

Slipping carefully along the edge of the crag, he got within a quarter of a mile of the sheep unobserved, but could see no way of approaching nearer without attracting the sentinel's attention. Nothing was visible but that big fellow on the rock, but the naturalist was confident that the whole band was somewhere in the immediate neighbourhood. Ascending a tree, he beheld a rare sight. In a sheltered cove below the crag on which stood the watcher was a band of seventeen big rams. Their fine proportions, their enormous curving horns, and their apparent freedom from all danger set the blood bounding through the veins of the hunter as he took in the situation from the top of that spruce tree.

For a full half-hour Dyche watched the animals from his tree-top, and then he began to plan a way of getting at them. Every foot of ground for five hundred yards in every direction from the sentinel's post was as bare as a floor, and there was little encouragement offered the naturalist. The old ram stood like a carved statue, his only movement being the turning of his head from one point of the compass to another. For several minutes he would gaze in one direction intently, and then jerk his head around and look another way, but he was so far from the hunter that the latter was unable to make out when he was looking in the direction of the naturalist and when he was turned the other way. At last Dyche determined to chance it and crawl up towards the crag when he thought the ram was looking from him.

The wind was in the hunter's favour, and had it not been for the old ram on the rock the stalking of the band would have been a very simple matter. As it was, the only feasible plan appeared to be to crawl over the grassy slope from the edge of the timber, keeping to the right of the cove in which the sheep were feeding. Hatchet and belt were left at the foot of the tree, and the campaign began.

Carefully crawling about fifty yards into the open space, the head of the sentinel came in sight outlined against the sky. The head was plainly to be seen, but the question to be solved was which way the animal was looking. After watching the horns for some time, Dyche decided that the ram was looking away from him. He began pushing himself along, watching those big horns all the while and stopping at the slightest movement of the sheep's head. This continued until noon, and the hunter was just beginning to congratulate himself on the success of his plan, when all his calculations were upset by the appearance of a second pair of horns on the crag.

As the new-comer seemed to be looking directly at him and as the sheep appeared to be moving around as if about to leave, Dyche feared that he had been discovered and that the band was preparing to make a break for some other mountain. The second sheep, however, lay down on the top of the rock, and the first sentinel went below to feed with the rest of the band. The naturalist pushed on, depending now on the movements of the head of the reclining sheep. Between one and two o'clock two more sheep made their appearance on the crag, and all three moved about, but finally two of them went below, leaving one watcher on the top.

Worming his way along, he reached a small water-worn gutter on the side of the slope. Crawling down this until it joined a larger one, he made good headway up the slope until he reached the head of the gutter. Fifty feet to his left was a string of low bushes, and by edging, inch by inch, across the intervening space, he soon had these for cover. Fifty yards more would take him to the edge of the cove where the band was feeding in fancied security. An old scrubby evergreen bush and a few irregularities in the ground afforded partial cover, and as Dyche was weary of his hours of crawling, he made for the bush with the hope of getting a little rest. By inches he worked his way, and after five hours' crawling he was at the edge of the cove, sheltered by a small scrubby tree.

Covering the crown of his cap with twigs from the tree, he slowly lifted his head and peered into the cove. In that single glimpse he felt repaid for all

HOW THE SENTINEL FELL.

the toil which he had undergone. Not over seventy-five yards away was a band of sixteen as fine rams as man ever set eyes on. He could hardly believe that he was awake, for there, in plain sight, were the sheep, some lying down chewing their cuds, others feeding, while still others were walking aimlessly about the cove. Now the naturalist began examining each individual member of the band for the purpose of selecting the best specimen.

There stood a monster, but his colour was a little too light. Ah, there is a grand fellow! He is the one. But no, his horns are blunted at the points. There is the right one. What magnificent horns! What a beautiful chestnut colour! He is the one to adorn the naturalist's exhibit at the World's Fair. The gun is carefully trained on the animal and almost fired, when a glance is given to the one on the top of the crag. He is evidently the patriarch of the band. What a beauty! Perfect in size, shape, and colour, with immense horns. But he is fully two hundred yards away. Shall the chances be taken? The question was soon solved, for Dyche made up his mind to have that leader if he never shot another sheep. The band was close enough to give him a second shot before it got out of range, and he trained the Remington on the sentinel. The old guard stood broadside to the hunter and presented a fine target. A puff of smoke, a loud report, and then there was the sound of a mighty rushing and scrambling of hoofs in the cove. Hurriedly slipping a fresh cartridge into the gun, the naturalist ran to the edge of the cove, but just got a glimpse of the band disappearing over the slope towards the woods. He was too late for another shot, as the last of the band went into the timber as the gun was raised.

Turning his attention to the old ram at which he had fired, Dyche hastened, as rapidly as the lay of the ground would permit, to the top of the crag. He was sure that he had not missed, but when he reached the spot not a sign of the ram was to be seen. Not a drop of blood, not a hair was found which would indicate that a wounded sheep had ever stood on that rock. The hunter's disappointment was almost too great to be borne. He had left a sure shot in a vain attempt to accomplish too much, and had lost the best opportunity he had ever had to secure a fine specimen.

Making a circuit of the crag, he saw where the ram had bounded away towards the woods. The tracks were plain, but not a drop of blood was to be seen anywhere along the trail. With a feeling of disappointment that almost amounted to despair, Dyche followed the trail mechanically. But he knew there was no possible hope of overtaking that band. Slowly he followed the tracks down the slope until he found where they joined those of the main band, and then he could see the broad trail where the fleeing sheep had ploughed up the ground in their mad leaps down the declivity. Two days' hunting and six hours of most wearisome crawling had been wasted, all because he was not satisfied with what was in his grasp, but must reach out for the unattainable.

With anything but pleasant thoughts the naturalist followed the trail of the fleeing animals through the park. Clear across the park he could see the fresh dirt which had been thrown up in the flight, but no sheep was in sight. He stood looking at the distant mountain-tops for a few minutes, debating whether it would be wise to follow the band, and then decided to return to camp and lament his folly. Just as he turned in the direction where old Charlie had been left feeding, a moving object in the edge of the timber caused him to throw up his gun. As he did so an old ram walked slowly out into the opening and stopped in plain view with its head down and then sank slowly to the ground. Dyche walked to within fifty yards of the animal, but it never moved, and approaching nearer, the naturalist saw that the ram at which he had fired was dead at his feet. The ball had entered too high to hit the heart and too low to break the back, and the wound was of such a nature that all the blood ran into the cavity of the body, which accounted for the absence of blood on the rocks.

The revulsion of feelings from his deep disappointment was so great that the hunter felt literally refreshed, and proceeded to measure and skin his prize, after which he carried the load to old Charlie and made his way to camp. He longed for his companions now, to share with him the joy of the prize, but he was too tired to waste time in regretting their absence, and was soon sound asleep after his hard day's work.

Dyche now had plenty of rams, but he needed ewes and lambs to complete his collection. Several days were spent in fruitless stalks after bands of sheep, and then it was decided to go farther north to another range, where possibly the ewes and lambs could be found. The specimens were cached, and the hunters started through deep cañons and over rocky crags, going two days in this way until they reached the wildest country they had yet seen. During the day **Dyche's** watch stopped, and in order to get the time two strings were hung in line with the north star. At noon next day, when the sun threw the shadow of one string on the other, the time-piece was set at twelve, and in honor of the event the place was named Camp North Star.

A band of sheep was seen that evening on the top of a high mountain, and next day Dyche saw a band of fourteen and made a long and tedious stalk to get near them. The animals were feeding in a patch of timber, and the hunter could get no closer than within two hundred yards of the band. The sheep were in a patch of timber on the side of the mountain, and the naturalist sat patiently waiting for them to feed towards him. They appeared to be travelling slowly his way, but he could only get occasional glimpses of them through the woods.

Suddenly he heard a rushing sound, and looking from his hiding-place he saw a small band of sheep flying up the mountain, while the others were making as good headway in another direction. The wind had changed, and they had got scent of the hunter and were thoroughly frightened. Well knowing the futility of a pursuit, Dyche made his way back to camp, killing a white-backed woodchuck and a grouse as he went.

A cold wind was blowing from the northwest and snow-flakes were filling the air next morning, but Dyche was determined to secure a lamb and ewe if possible. Saddling old Charlie, he rode five miles west. About noon he saw two old ewes and a yearling. Jumping from the horse, he stalked the little band which was working his way. They fed on towards him until they were within sixty yards of his hiding-place. Concluding that he could kill the yearling with a load of shot from the shot-gun and follow it up by killing a ewe with the rifle, the naturalist opened fire and saw the yearling running at full speed towards the timber as if nothing had hit it. Snatching up the rifle he sent a ball after it, dropping it at the edge of the timber. The ewes, of course, were out of sight, and this incident caused Dyche ever after to discard the shot-gun when hunting big game.

As supplies were running short the specimens were taken to Thorp's ranch, and the hunter outfitted for another hunt after ewes and lambs. The trip to the ranch and back to Camp North Star was made without incident, and on the morning after the return Dyche took an early start, and by sunrise was on the flat top of a mountain where he had seen so many indications of game.

About ten o'clock a spot was reached where the numerous signs showed that a large band of sheep had been feeding recently. On the south side of the mountain the wind and rain of summer and the frosts of winter for centuries had been breaking off masses of rock, which had accumulated in a sheltered nook. The sheep had pawed out this soft, crumbling rock and made beds, which were lined with gravel and smooth stones and were dry, showing that they had been occupied very recently. There were over seventy of these beds, and Dyche began to work very carefully, for he was confident that he was near a large band. About half a mile from the beds he reached the edge of the flat top of the mountain, and crawling to the edge he peered over. A great, undulating slope stretched from the mountain-top to the woods on the other side, and this was scanned closely, until at last the naturalist made out moving objects below. They were sheep, and over thirty were in sight at one time. The wind was in the hunter's favour, but the country was bad for stalking, as the animals were feeding over a large open space and were scattered in small bands.

Making a long circuit, Dyche went down a cañon and approached the game from another direction. From his new position he could count fifty-seven sheep, but the nearest was three hundred yards away, and the naturalist could see no way of getting nearer without being seen. He took another two-mile circuit and again approached the timber. From its edge he crawled up a small ravine, fearing each minute that some of the animals would take alarm and scatter the band. The head of the ravine was finally reached, and lifting his head carefully he saw five sheep within a hundred yards. Two old ewes and three small bucks, probably yearlings, were feeding within easy range, but Dyche wanted a lamb, and hesitated whether to shoot at the sheep in sight or wait for a better opportunity.

Deciding to try again he began a retrograde movement, although he ran the risk of frightening away the whole band. He felt, however, that if he missed a lamb this time he would have great difficulty in getting so close to one again. Crawling slowly up another ravine, he got sight of an old ewe and a lamb. They were just what he wanted, but they were over two hundred yards off and there was no possible way of getting **closer.** He determined to let them go, and crawled back to the woods **and went up another ravine.** He had crawled about **two hundred yards** when two sheep fed along **the head of** the ravine. The naturalist flattened himself **to the ground,** fearful that he had been seen. The animals fed in almost the same place for fully an hour, and the hunter was just beginning to feel that he would have to move even at the risk of scaring away the whole band, when they went around a knoll.

Crawling to the head of the ravine Dyche carefully lifted his head, but there was nothing in sight. Waiting a few minutes he crawled out on the grassy slope, hiding himself as much as possible, flattening himself to the ground. His movement was evidently seen by some watchful sheep, for he suddenly heard a tremendous rushing of the frightened animals, and jumping to his feet saw sheep everywhere going at full speed. The Winchester was trained at the flying animals, and an old ewe received the first shot. Without waiting to see the effect of the shot, another ball was sent after a lamb that was just going over a knoll. The ewe made a few jumps and dropped, and fifty yards over the knoll the lamb lay dead.

The bullet had struck him in the hip and ranged forwards to the neck, not coming out.

All this stalking had occupied much time, and it was now between three and four o'clock in the afternoon and camp was a long way off. An hour was spent taking notes and measurements and skinning the animals, and then Dyche started for camp, carrying the load of skins, skulls, and a portion of the lamb. The way was long and rough, and finally he was compelled to cache the greater portion of his load and proceed with his quarter of lamb.

Next morning it was snowing heavily, and the northwest wind cut to the bone. Dyche waited for a cessation of the storm, but finding that there were no indications of a lull he saddled Slim Jim, the meanest and fattest horse in camp, and started after his specimens. The snow had changed the appearance of the country so much that he could hardly make out the landmarks which he had noticed the day before, but just as he was about to give up the search he saw a moving object among the trees. Suddenly an old ram came into view at the edge of the timber. Jumping from his horse, Dyche began to unfasten his gun and the ram started for the timber, climbing up the rocky ledge. By the time the gun was out of its fastenings the ram was at the top of the ledge, but here he made a mistake. He stopped to look back. Just at that moment Dyche dropped on one knee and sent a bullet after him. As the gun cracked the sheep jumped away, and Slim Jim did likewise. Dyche spent fifteen minutes catching the horse, and then went to look for the sheep, which he found just over the ledge, stone dead.

Very little time was spent in the measurements and skinning, and Dyche started to camp with the load, intending to leave the ewe and lamb for next day; he knew that nothing would touch them, for two handkerchiefs floated over them as danger-signals. While wandering around trying to find the way to camp he reached the spot where he had killed the ewe and lamb, and thus getting his bearings soon found where he had cached the skeletons and skins. Placing these on Jim's back, the whole load was taken to camp.

The naturalist now had a complete collection of sheep of all ages and kinds which made up a complete family, and he concluded that he had had enough sheep-hunting. In addition to the sheep, he had many smaller specimens and several fine mule buck skins and skeletons. Several days were spent in getting out of the country. Horns and skeletons are not easily carried on horses, for the rough and rocky trails are liable to cause accidents which will break the fragile bones. It was after dark when camp was made the first night, and it took until nine o'clock next morning to get the packs arranged again. Everything was piled up in such a way that the place was named Camp Confusion.

About noon the trail led through a patch of blueberries, and here the horses were unpacked and the hunters regaled themselves on the berries. These were of two kinds: a large variety with a whitish bloom on the berry and a small black one. The large ones were full of meat, while the small ones were very juicy, and a judicious combination made a very palatable dish.

Dyche and his companion had now been living on a meat diet for ten days, and they were beginning to feel the effects of it. As they were within a day's journey of Thorp's ranch, a pint of navy beans, which had been held in reserve in case of sickness, was brought out and made into soup. So well did the supper please the naturalist that the camp was at once christened Camp Bean Soup in honour of the occasion. Breakfast was made from the rest of the soup next morning, and late in the afternoon they reached Thorp's ranch, tired and hungry.

Thorp's garden, full of ripe vegetables, had a charm for the hungry hunters, and Dyche regaled himself with tomatoes from the vines. Letters from home were waiting for him by the dozen. The regular rate of postage was twenty-five cents, but Loomis, the postmaster, bunched the lot at ten cents each.

Several days were spent in packing and storing specimens, but Dyche was not yet ready to leave the country. He knew that Rocky Mountain goats and caribou were to be found somewhere in this vicinity, and he meant to have some of them if possible. Many were the stories told him about the mysterious Kettle River country. There were hundreds of miles of unbroken forests where foot of white man had never trod, where game in vast numbers was so tame that deer would not run when man approached, and where wolves roamed in ferocious bands. While Dyche was a little sceptical about these stories, he thought there must be something in them. Caribou and goats were the specimens that he now wanted. Indians told of great droves of caribou which wandered hundreds of miles up in the woods at the headwaters of Kettle River, and the naturalist felt that he could go where an Indian could, and he determined to make the journey.

CHAPTER XI.

On Kettle River—Okonagan Smith and his Lonely Ranch—The Great Northern Boundary—Trials and Tribulations—"Fool Hens"—Through Fallen Trees—An Arduous Journey.

ONE hundred miles up the river was a cabin built by three noted trappers and mountaineers—Farrell, McLaughlin, and Dore. Beyond this cabin was an unexplored wilderness where no white man was ever known to have been and where few Indians had ever wandered. An unbroken forest stretched away hundreds of miles to the far north to the foot of the white-topped mountains, on whose slopes vast herds of deer and caribou were supposed to roam, and where bears, wolves, and mountain lions were thought to be in countless numbers. The white crags and peaks of the mountains were supposed to be the home of goats and sheep. All this was surmise, however, for man had never been there to report. Caribou had come down as far as the trappers' cabin during the coldest weather of the winter before, and five had been killed and many seen near the camp. Immense droves of gray wolves had ranged about the cabin, while wolverines and bears had left many tracks.

All these stories indicated that the trip would be interesting, to say the least; but Dyche had heard too much of this kind of talk to be deterred from his undertaking, and after two days of rest at the trading post he was ready to start into the unknown land.

Twenty miles of tedious travel over a dry and dusty road carried them to Okonagan Smith's ranch, on the shore of the lake which had given its name to the settler. For thirty years this man had been cultivating a piece of land there, and he still has the country all to himself. His ranch extended into the lake in such a manner that it could easily be irrigated, and it held a fine bearing peach and apple orchard. Here, with his Indian wife and two half-breed daughters, he was happy and contented away from the cares of civilisation.

After a bad night on account of the kicking and squealing horses, an early start from the ranch was made. The route lay eastward, over a country where there was plenty of wood, water, and grass, and Dyche felt that that trip was to be one of pleasure instead of the hardships which he had previously undergone. As the sequel will show, his calculations were wrong. While passing along over easy trails through the timber, the naturalists came into an opening where the trees had been cut down and a broad way cleaned off as if a right of way for a railroad had been cleared through the forest. The work had evidently been done several years before, as the stumps were old. The broad path could be seen stretching away for miles in each direction, going right over mountain and through valley, never deviating from a straight line. The strange appearance greatly puzzled the hunter and his companion, and it was not until they reached a small settlement on Rock Creek that the mystery was cleared. They had seen the dividing line between the dominions of England and the United States.

That night they were compelled to camp in the dark, with neither water nor grass, and they christened the place Camp Necessity. When morning came they found that they were near an old Indian camp. On all sides bones of deer were found, and a stack of antlers twenty feet high had been piled up between two trees. They travelled until nine o'clock before they found water and grass. All day long their way lay through a dense forest. No life was to be seen except vegetable life, and not a sound was to be heard except the wind in the tree-tops. After a hard struggle through the fallen timber an open spot was reached on the river-bank late in the evening. A beaver dam showed its top just above the water, and the trees on the bank gave evidences of the work of the industrious animals. One tree which had been cut had a diameter of over four feet, and from this the camp received the name of Camp Beaver Tree.

A light rain, the first of the season, fell during the night, and next morning they started early, trying to follow the old trail of the trappers. About the middle of the afternoon the cabin, on the banks of a little stream which emptied into Kettle River, was reached, but no stop was made. About four miles farther up the river an opening was found where there was grass and water, and here camp was made for the night. The stillness of the dense woods was simply awful. The only sound that broke the silence was the dismal hooting of a number of owls which made night hideous with their peculiar cry. The uncanny sound produced such a nervousness that sleep was much interrupted, and after naming the place Camp Owl Hoot the hunters made haste away from the spot. A tall tree was climbed and the lay of the land noticed, and the northward way resumed about daylight.

An almost impenetrable forest was now before them. No trail was to be found. The rivulet was dry, and the bed of this was followed until it ended at the foot of a ridge. On the ridge a grassy spot was found where a few deer-tracks gave the first indication of animal life. In every direction from this spot dense thickets of poles stood in such confusion that it was necessary to cut a way through. This labourious proceeding was continued until late in the evening, when a small lake was reached and camp made for the night. The lake was alive with waterbirds, and a fat mallard duck made a supper for the hunters. All seemed peaceful when the tired naturalists crawled into their sleeping-bags, but scarcely had they composed themselves when a series of shrieks and yells, which appeared to come from the vicinity of the lake, sent them bounding from their beds to their guns. The sound died away, and after waiting in suspense for some time they again sought rest. They slept soundly until just at daylight, when again the unearthly cries broke the stillness, and the hunters hastened to their feet. An investigation showed that a flock of loons had been feeding on the lake near the camp, and it was their cries that caused the alarm.

No sign of a trail could be found, and all day the cutting of poles was continued. Night found the hunters still in the midst of thick woods, and Dyche, finding a caribou horn, took it as an omen of good luck and made camp on the spot.

All next day was spent in a similar manner. The party was still headed up Kettle River, and penetrated the dense mass of poles and fallen trees. They were in a veritable *terra incognita*. No hatchet or axe had ever been used in the whole expanse of country. No sign of the presence of man was to be found anywhere. It was hard and lonesome travelling, and forage for the horses was always uncertain, but there was a fascination about it. They never knew at what moment they might meet with some exciting adventure.

Just as night was coming on they saw a flock of Canadian grouse, which were so tame that six of them were killed with sticks and stones. These beautiful birds are known in the country as "fool hens," on account of their tameness. Camp Fool Hen was christened and supper made of the birds. The cocks are a mottled gray with black breasts. A small fiery red comb stands up above the eyes, and usually they are strutting around like miniature turkey-gobblers. When approached they would run a few feet out of the way or fly to the lowest branches of the nearest tree and sit until they were knocked over with sticks. Over thirty of them were killed on this trip up Kettle River with sticks and stones. Dyche found four in an open space, and with a pole twelve feet long killed all of them.

All next day was spent in a wearisome march over fallen logs and through pole thickets, until the whole party, horses and all, were almost worn out. All day long the horses had been led through the wilderness, jumping over logs and crawling through narrow places until their legs were covered with scratches and bruises. Old Charlie would try anything in the shape of a log, and would jump any not higher than his breast. Billie was of different mould, and would stand and shake his head at every obstruction. If he did not jump when the head-shaking was finished, that was the end of it, for no power on earth could make him go over, and he had to be led around or a path cut through for him. Camp was made in the densest woods where a little water seeped through the moss, and as a chickadee flew near and enlivened the place with his chirp, the place was christened Camp Chickadee.

Next morning Dyche climbed to the top of the tallest tree in the vicinity and endeavoured to make out the way through the forest. To the north rose the white-topped mountains, seemingly only about ten miles away, and this sight so inspired him that the party went forwards with better heart. For about a mile the way led through comparatively open timber, and then an immense spruce forest was entered. A fire had raged through this some time before and left a mass of fallen logs that was almost impenetrable. Huge logs lay in greatest confusion, compelling the travellers to turn and twist in every conceivable direction, over and under, until they had to stop from sheer exhaustion. The horses were jumped over logs until they could jump no more. The hunters chopped logs until it was almost impossible to lift an axe. Then they came to a stand-still. A little prospecting discovered water and grass in a swampy place near by, and an hour was consumed in getting the horses to the spot. It was late in the afternoon, and camp was made in one of the wildest places ever visited by man. The swamp was full of high and low bush blueberries, and a quart of these supplied the supper, while the sparse grass made a meal for the horses. One day of rest was taken here, for while it was not a good place to camp, it was better than the continuous travel through the wilderness of poles and fallen timber.

The density of the forest was so great that the white tent could not be seen fifty yards away, and to leave camp was a dangerous undertaking. One person was compelled to remain at the tent all the time to give necessary signals when the other went out to look for game or to spy out the way. Bear, deer, and caribou tracks were seen in the swamp, but it was impossible to hunt them in the dense woods. One deer came near the camp and was shot for meat.

A careful examination of the country showed the futility of an attempt at farther progress in that direction, and the back track was taken for five or six miles, and then an attempt was made to go east. Seven miles in this direction were covered when night came on, and they were compelled to camp. Dyche started on a prospecting tour, blazing his way, in an attempt to find some way out of the country, but he gave it up in despair. Naming the place Camp Last Attempt, next morning they turned south and travelled until they found a place where the ground was moist. A hole was dug, and the seepage gave enough water to fill a quart-cup. So careful were the hunters with this that they dipped it from the hole with a spoon for fear of losing a drop. The horses were compelled to go without water until Camp Chickadee was reached next day.

Resolving on one more effort next day, a northeasterly direction was taken with the hope of leaving the worst of the swamp to the west. After a hard day's travel a small opening in the woods was found where there was water and grass. Camp was made, and Dyche was soon at the top of a tall tree looking over the country. The bald tops of the mountains looked invitingly near, and it was determined to get to them if possible. There was sufficient grass to last the horses two days, and it was decided to leave them here and proceed on foot to the base of the mountain. Dyche carried the axe, and the two men started due north, blazing a trail through the deep woods. They were compelled to zigzag through the timber, and while the blazes were frequent and large, there were none too many when they returned.

At three o'clock in the afternoon the base of the mountains was reached, and they hastened to the top, where many tracks of deer and caribou were seen. On all sides droves of deer were feeding on the rich grass which grew in profusion on the slopes. From the top of the mountain the country was examined. The apparently boundless forest through which they had come appeared to be a small grove compared with the forest which spread out before them. An hour was spent here, and then the hunters hastened down to the timber to find the way back to camp. At the edge of the timber they found the blazed trail which led to the little white tent miles away in the trackless forest. If the darkness should prevent them from finding this trail, then the search for the tent would be hopeless indeed.

Evening came on and then darkness, and then the blazes could no longer be seen. One of the hunters would stand by a blazed tree while the other went on until he found the next one, and thus they proceeded, almost groping their way, until half-past ten, when they came to the opening and found the tent. It was late next morning when they arose, and they at once came to the conclusion that the only thing left to do was to get back out of the country as soon as possible. It was evident that they could not hunt to advantage in such a jungle, and if they killed anything it would be almost impossible to get the specimens out of the country. The pack-horses were so weakened by lack of grass and water that they could carry nothing more than they had, and the specimens would be so torn that they would be useless. The back track was taken in a snow-storm, and the hunters were almost discouraged over their bad luck. The whole day was spent chopping a trail through the dense thicket with neither grass nor water.

They were compelled to camp without water, and

when they began to prepare supper they found that a hole had been torn in the gunny-sack in which was carried the meat, and the bacon and remaining venison had fallen out on the trail. Search was made for water, and a little was found a quarter of a mile away. Supper of oatmeal mush, biscuit, apple-sauce, and tea was a light diet for hungry men, but it was the best that now remained until they could find some game. Next morning Dyche hunted back three or four miles in hopes of finding the lost meat, but with no success. He killed a "fool hen," but all other animal life appeared to have deserted the forest. Even the few woodpeckers which flitted about the trees seemed so cowed by the stillness that they neither tapped the trees nor cried.

The silence of these forests was fairly appalling. Not a sound broke the quiet but the moaning of the trees as they rubbed against each other. It rained, and the water came through the dense tops as a fine mist. It snowed, and the flakes sifted down through the spruce boughs like finely ground flour. In this dreary lonesomeness the sound of the human voice appeared strange, and words were rarely spoken above a whisper.

In the course of the night the snow turned to rain, and next morning an easterly direction was taken. Down a cañon and up on a ridge their way led for two hours, when the top of the mountain was reached. From the tops of trees it was discovered that a flat country lay to the northeast, and the trail was turned in this direction. Again and again were trees climbed in hope of seeing something, and at last a number of lakes appeared still farther to the east. A straight trail was taken to this spot, and the lake was reached just at dark in the midst of a heavy snow-storm. The body of water was small and was full of beaver, from which fact it received the name of Beaver Lake.

After breakfast next morning a little prospecting was done, and it was found that Beaver Lake emptied into a larger lake. In the centre of the large lake was a small island, and it was named Island Lake. Camp was moved to the shore of Island Lake. Here was found a regular runway of animals. A path made by bears, deer, caribou, wolves, and wolverines led all around the lake. It was evidently the spot where game could be found in abundance, and was just the place for a long stay. But Dyche and his companion had a dispute over the direction of the points of the compass, and as the sun decided in favour of the naturalist, his guide had such a fit of sulks that he insisted upon leaving the country at once. It was a foolish move, but it would have been a worse one for Dyche to have attempted to remain there alone. The back trail was taken next morning through deep snow, under bushes which were bending with their load of the frozen crystals.

Out of meat and with provisions getting scarcer every day, the situation was desperate. While bean soup was considered quite a delicacy when properly made and when not eaten too often, it was found that it lacked flavour when it lacked bacon. "What is bean soup without bacon?" was an expression that became indelibly fixed in the mind of the naturalist.

All day was spent in travelling through the forest, and late in the evening camp was made about four miles above the old cabin, whence the trail led out to civilisation. A halt was made here for one day, and a fine two-point buck was killed, giving them welcome meat after their privations. At four o'clock in the afternoon of the next day the cabin was reached, and was found to be occupied by four prospectors, who were delighted over the "colours" which they had found in the creek. But indications of gold held no charm for the naturalist, and the way down the river was continued. The horses were in bad condition, and they travelled very slowly. During one of his side excursions after a straying horse, Dyche shot a coyote, and this, with a dozen or so skins of "fool hens," constituted the specimens secured in a trip of over a month in the wildest part of North America.

For two days they now travelled on a diet of venison, as their provisions were gone. A delicious repast of ripe peaches from Thorp's orchard freshened them up and a good supper renewed their energies so that they slept without care.

CHAPTER XII.

In the Cascades—A Forest Fire—After Rocky Mountain Goats—The First Shot—Down the Mountain—A Successful Hunter—A Night of Hardships—A Naturalist's Labours.

THREE days spent at Thorp's ranch preparing for an expedition after Rocky Mountain goats put the horses in fair condition for travelling, and they went along at a good rate on the first day. The whole country to the north and northwest was covered with smoke from the forest fires which had been started by Indians to bunch the deer. Hundreds of the animals were slaughtered and thousands of acres of the most magnificent timber land in the country burned over and ruined by these Indians. On the first day out from Thorp's ranch thirty-seven deer were seen. There was a camp of Indians near old Camp Victoria, and here Dyche stopped for a few minutes' talk with the redskins. Dozens of dogs greeted him as he approached the camp, and about a dozen bucks came out and stretched themselves on the ground near the naturalist, with the exclamation "How!"

As the red men could talk a little broken English, the hunter endeavoured to get information from them regarding game. The spokesman held up five fingers and said, "Mowwich," indicating that he had killed five deer. Then pointing to each of his companions in turn, he held up the number of fingers suitable to their achievements and repeated the word. Dyche asked about sheep and the Indian pointed to the mountains far to the west. He then asked about goats, or "white sheep." The Indians consulted a little and then pointed to the crags in the northwest and said, "White sheep high up." One Indian supplemented this information with the remark, "Me no hunt him," whereat the others laughed boisterously. Dyche declined the invitation to "come down," and rode on until a suitable camping-place was found.

Several days were now spent on the trail over mountains, through masses of rock and down timber, while the whole country was almost obscured by the smoke from fires which were raging on the other side of the range. Finally farther headway was prevented by the fire, and in a little valley, through which leaped a sparkling stream and along whose edges grew an abundance of grass, camp was made. A fence of poles behind them and the fire below prevented the horses from straying away. As usual, old Charlie would not go away from the tent, and it became necessary to lead him down to the stream to keep him away from the camp. He seemed to prefer the smell of the camp to the best grass on the range. During the whole time he was in the mountains he always remained within a few yards of the tent unless he was driven away.

Immediately after camp was made Dyche started on an exploring expedition. A light breeze was blowing the smoke in great clouds up the east side of the cañon, and Dyche climbed the west side. After hours of hard work he reached a projecting crag, where the full beauty and magnificence of the mountain fire burst on his gaze. Thousands upon thousands of veritable monarchs of the forest were being swallowed up in the dreadful conflagration. Ruin was spreading over miles of territory, simply that the lazy Indians might hunt more easily.

On the way down the mountain he noticed a bunch of white bristly hair attached to a jutting rock, and by the light of the camp-fire he decided that it had been scratched from the side of a Rocky Mountain goat. This convinced the naturalist that he was near the animals for which this trip had been made, and daylight next morning found him on his way up the side of the mountain.

Here were the steepest and most inaccessible of the steep and rugged Cascades. Long ages of frost and sunshine had torn great masses of rock from the sides of the projecting crags, which had plunged to the bottom, ploughing immense furrows down the mountain's sides and piling up in a confused jumble at the base. From the almost perpendicular sides of the precipices projected here and there shelves or benches of rock, to which clung stunted and dwarfed growths of pines and spruce, while zigzagging up from one bench to another were little gulches or chasms which gave the explorer an opportunity to reach the top after the expenditure of great labour. Two hours of hard work placed Dyche about two-thirds of the way up the mountain, and here he found a shelf running apparently around the face of the crags. Following along this bench he **reached a point** overlooking the entire valley below, and here he found, in the crumbling rock **and débris, the** bed **of** a goat. It **was** similar to **those** which the sheep had made, but sticking to the sides could be seen the long bristly white hairs of the goat.

Here evidently the old billy had selected a spot in which to chew his cud, for from this **point** he could view the whole country **at his** leisure. **A short** distance farther on three more beds were found, and many tracks indicated that this bench was the favourite resting-place **for** goats. While proceeding cautiously, Dyche was suddenly brought to a stand-still by the noise of a stone rolling down the mountain ahead of him. Well knowing that when stones move **some** animal life must be near, he stopped and listened until the noise was repeated, and then he ran to the end of the bench, about a hundred yards beyond. The increasing noise showed that some animal was hastening down the mountain.

Slipping off his shoes and depositing his camera with them, the naturalist started on **a noiseless run** towards the **place whence came** the **sound**. At this point the formation of the mountain was peculiar, in that there were two shelves or benches, one a short distance above the other. These were connected by a series of gulches, thus leaving a number of jutting crags extending from the upper bench. The animal reached the end of the lower bench as the hunter arrived at the end of the upper, and now began a race back. As Dyche reached the crag he saw some white, shaggy object just disappearing around the rock below him. The first thought that flashed across the mind of the naturalist was that it answered the description of a polar bear, and then he realised that he had seen a goat. The animal was skulking along on the lower bench at a fair gait, and Dyche turned and ran back along the upper ledge as fast and silently as possible. He reached the head of the first gulch just in time to see the goat go behind the next crag, and then made another run of about forty yards to the next opening, where he got another sight of it. Dyche was running up and down the mountain, while the animal was going along on a level. Almost exhausted, and breathless from running so violently in the high altitude, his lungs working to their full capacity, perspiration pouring down his face, and his heart throbbing as if it would burst, the hunter pressed on.

Another run and a rest and he could hear the stones rolling ahead of him. Again he ran fifty yards, calling into play all his surplus energy, and waited on the point of a crag, but again he was just too late. He could hear the animal still ahead of him. A run of a hundred yards at full speed placed the hunter on a crag just in time to see the goat disappear about seventy yards away. This sight was sufficient to nerve the naturalist for one more effort. A grand spurt was made for about twenty-five yards up hill, and the head of the cove was reached. Dyche stopped, gasping for breath and almost exhausted, but stood stock-still and listened. Suddenly a big white object sprang upon a crag in front of and below him, not over seventy-five yards away. It was a large goat, the animal which had given him such a chase. Wiping the perspiration from his eyes, the naturalist drew down the Winchester and compelled his muscles to be steady long enough to take good aim, notwithstanding the fact that he was almost sinking from exhaustion. Logs and rocks obscured the goat, so that Dyche was compelled to stand on tiptoe to get sight at the top of the goat's back.

With the report of the gun the animal disappeared in a cloud of dust which rose from the spot below where it had been standing. Down the cañon rushed the excited naturalist after the goat. Rolling and tumbling, he went down the gulch to the place where the goat had been, but it was gone. A cloud of dust, however, showed that the animal was also rolling down the declivity ahead of him, and without waiting to figure on the consequences the hunter dashed after his prize. The melting snow had left a mass of débris in the gulch which had been ground to dust and which was now dry and light. The struggles of the goat sent this dust rolling in clouds back to where the naturalist was following and almost choked him. Slipping and falling he hurried on, and at last saw the shaggy brute lying on its back by the side of a log. As it stood on the crag it had looked like a small albino buffalo, and when dead the resemblance was almost perfect. The ball had gone through the top of the back just above the spine and had paralyzed the fore-legs. The goat was making ineffectual struggles to get on its feet. Dyche rolled down to within a few feet of the animal and lay there so exhausted that he could not move.

It was well that the goat was disabled, for the hunter could have done nothing to have stopped it had it recovered sufficiently to start away. Thus he lay for full five minutes, gasping, before he could recover sufficiently to attend to his game. Regaining his breath and strength by degrees, the naturalist finally began taking notes, and having dispatched the goat with his knife, he walked back to the place where he had left his camera. He photographed the animal just as it had fallen. Three hours were now spent in measuring and skinning the goat, and at three o'clock in the afternoon the hunter started for camp at the foot of a steep mountain.

The package of skin and bones was about as heavy a load as Dyche could well carry, and to add to his discomfort he was terribly thirsty. He had had nothing to drink since he left camp early in the morning, and his great exertions had caused such violent perspiration that his system was depleted of water. At four o'clock he had gone about half a mile, and found himself on a shelf from which there appeared no possible means of farther descent. A lower bench was finally reached, but no water could be found. While searching here for a path another goat was seen feeding on a rocky ridge across a small cañon about four hundred yards away.

Removing his shoes, the naturalist began a stalk, but soon found that he could not get nearer than three hundred yards in that direction, so he crawled back and descended to a lower level. Slipping along the ridge, he reached a spot which he supposed would be within easy range of the goat. He peered round the edge of the crag, but the animal had disappeared. The goat, however, came from behind a rock and stood about a hundred yards away. He was across a cañon, but the naturalist rested the Winchester on a rock and after steady aim fired. The goat plunged head foremost over the cliff into the cañon below. Dyche hastened back after his shoes and load of skins, and went after his second prize, which he found at the foot of the precipice.

The naturalist's thirst was so overpowering that he determined to drink the blood of the goat, but when he cut the animal open the strong musky odour was too much for him, and he gave up the idea. Finding that little progress could be made down the mountains with his load, he now determined to cache his specimens and come back for them next day. A tree was stripped of its branches with the exception of a bunch at the top, and to this he tied the bundle of skins and bones, and to make it even more conspicuous he tied his camera to the bare portion of the trunk. The descent to the bottom of the mountain was now made, and here he piled up stones to mark the spot, and placed three piles of small stones on a log. It was dark now.

Fully satisfying himself that he would have no difficulty in finding the place where he had come down the mountain, Dyche now set himself to the task of reaching the river in order that he might quench the almost overpowering thirst. The river was near, but the great masses of stone which had crumbled from the mountain presented an almost impassable barrier. At last the edge of the river was reached, but he found another difficulty. Great piles of rocks barred the way, and the steep declivity precluded all possibility of getting down. The sound of the running and babbling water as it dashed over the stones in the river-bed was maddening. Dyche struggled along, over bowlders and under logs, and through bushes which held him back as if leagued with the powers of darkness to restrain him in his efforts. Darkness came on and still he struggled on, fearing all the while that he would fall into some hole in the rocks from which he could never get out. Feeling his way, he finally emerged, but he was worse off than before, for here the fallen timber, overgrown with bushes and briers, was so heavy that it made a dense jungle. At times he would run against a great rock that stood higher than his head, and then he would get into a tangle of briers so thick that minutes were spent in getting through.

His exhaustion was now so great and his thirst so maddening that the naturalist tore his way through the briers and underbrush, unmindful of the cuts and scratches. He was in a frightful condition, and felt that his brain was turning from his tormenting thirst. A patch of "devil's walking-sticks," a shrub whose curved thorns hold all they catch, barred his way, and he felt as if Satan himself had lent a hand to keep him from getting water. At last, however, he reached the river. A log reaching from the bank gave him access to the water, and after washing his burning face and hands he began the work of quenching his thirst. He drank slowly and finally felt that he had sufficient. But as he started from the water he felt the thirst again and drank again and again.

Reaching the top of the bank, the naturalist sat down to consider his situation. He had promised himself that when he reached water he would camp for the night, but now he felt so refreshed that he thought he would get to the tent and have a good rest and breakfast before he started again for his specimens. By this time he was again very thirsty, and he made another trip to the water before he started up the river. He knew the camp was above him, but just how far he could not tell. An hour or more was spent in getting through the underbrush, and then he was compelled to get another drink. He now determined to make one more effort, and if he failed he would give up for the night. Stumbling along over rocks and poles, he soon found himself in a wilderness of huge logs. It was so dark that he ran into the trees before he was aware of their vicinity. Feeling his way along, stumbling and falling, with bruised and bleeding limbs, he was indeed a pitiable object, but he pushed on and on. Finally he sat on a log and gave up. He dozed and nodded from exhaustion, when suddenly he heard a shot away up the cañon. It was evidently the guide making signals to bring him into camp. Giving the answer he now pushed on, wondering why the shots sounded so far away. Once again the shots were exchanged, and from the sound he judged the camp must be fully five miles away.

Striking a match, he looked at his watch and found that it was just eleven o'clock.

His hands and feet were burning and he was so dead tired that when he sat down to rest he would almost instantly drop to sleep, to be awakened by almost falling from his seat. Then he would push on again. The rushing of the water over the bowlders, the weird sounds of the forest, the roaring of the fires which raged on the opposite side of the cañon combined to make night hideous and still farther depress him in his terrible condition; but he pushed on, determined to reach camp, now that he knew the direction of it.

The extreme exhaustion now began to tell most strangely on Dyche's brain. Odd fancies and queer hallucinations flashed through his mind, and thoughts that under ordinary circumstances would have appeared foolish now had serious consideration. At last he reached a little opening in the jungle and found himself in a small park. He had just begun to breathe freer, when there arose at his feet some huge animal which made two or three bounds away and then stood stock-still. To say that the naturalist was scared is putting it mildly. He lost entirely the little of his self-control which had remained after the awful experience of the night in the jungle. Tears involuntarily came into his eyes, his blood seemed to stand still, while chills began at his feet and crept all over his body, up and down. Frightened? He was so frightened that he felt that he should go insane unless something happened to relieve the awful strain.

He threw his gun forwards, of course, at the first movement of the animal, but the strange silence puzzled him. **Not a** sound, not a movement was made **by** the big beast. Dyche thought he would light a match, but gave up the idea for fear he would be attacked when the brute saw what a puny thing he was. He moved a few steps forwards, making as much noise as possible, but the animal did **not move.** Again he advanced, shuffling his **feet,** and the strange object moved only **a step or two** and again **stood** still. **Dyche knew** of no animal **as** large as this which would act in **that way,** unless **it was a grizzly** bear which had never seen a **man.** His next impulse was to shout, **for he** knew that the human voice had a wonderful effect upon wild animals. Acting **on** this idea he shrieked and shouted, but not a sound came from the place where the strange beast stood. He called his guide, he called all the names he could think of, but not a motion was made. Then he shouted again and called his old horse Charlie. **The** result was most startling. Old Charlie answered with a whinny from the very spot where the **strange beast** stood. Another call brought the horse to him, and then Dyche cried from revulsion of feeling. He petted the horse, and then fell to wondering what could have brought him so far from camp. It was the horse that had to be driven away to grass. Knowing that there must be a trail over which the horse had come, Dyche took hold of his tail and endeavoured to drive him back to the tent.

Charlie went forwards **a few** paces and then stopped. Dyche clucked at him and struck him, but the old fellow would only wander around in a circle. At last the horse brushed past a pole on which hung a gunny-sack. Dyche remembered noticing this as he came into the cañon, and now he began to wonder how far that pole was from camp. He struck Charlie again and made him go forwards. Suddenly there loomed up beside him some big white object which he took for a rock, but on feeling it found that it was the tent. Charlie had not broken his record, but was right in camp. In a few minutes a fire was burning brightly and supper was under way. It was after one o'clock in the morning, and the guide was not there. The shots had evidently been fired far up the cañon by the man, who was also lost. He returned to camp next day, having lain out on the mountain all night.

By half-past two Dyche finished his supper, or breakfast, and bath, and rolled into his sleeping-bag, feeling that "all's well that ends well." But the troubles of the night were not over. Just as he began to doze he felt something run across his bed, and then heard a dragging noise on the floor of the tent. A mountain rat was exploring the place. After standing it as long as h could, the naturalist got up and set a trap for the intruder. He had scarcely got back to bed when the trap snapped and the rat squealed. This noise had to be stopped, so Dyche got up and killed the rat, setting the trap again. The minute he touched the bed another marauder was caught, and then until morning the trap was kept busy and a family of seven rats were killed. When the last one was gone it was broad daylight, and not a wink of sleep had Dyche had. For once in his life the episodes of the night were too much for the naturalist's nerves and he could not sleep. He lay for two hours thinking of the specimens away up the mountain which must be brought into camp and taken care of.

The previous day had been nineteen hours long and had been supplemented by a sleepless night, and Dyche felt little like again climbing the mountain; but he knew that if he wanted to preserve his goat-skins he must bring them into camp. His feet were bruised, his hands and face scratched and sore from the briers and "devil's walking-sticks," and his whole body ached from the extraordinary exertions of the previous day; but he started after his specimens. Not caring to repeat his experience of thirst, he took a powder-can of water with him. Following down the river over the bowlders and through the thickets, he finally found the piles of stones which marked the spot where he had descended the mountain.

But something was evidently wrong. Hours were spent searching for the skins, but not a sign of them or the tree which had been so plainly marked could be seen. Up and down he climbed, searching in every direction, and at last, after going away off to one side, he found the place where he had killed the old goat, and then following his trail he found the cached specimens. He then found that in going down the mountain the evening before he had gone over a mile to one side before he reached the bottom of the cliff. It was half-past three in the afternoon when he reached the tree, and it was three hours later when he arrived at the foot of the mountain with his load. He dreaded the hard climb over the rocks to camp, but determined to do as much as possible. Leaving the skin of one goat and the skulls hanging to a rock, he started with the skin of the old billy, which weighed thirty-eight pounds, and managed to reach camp after eight o'clock. He was soon asleep, after nearly forty hours of sleeplessness and terrible exertion.

Next morning Dyche's knees were so sore and stiff that he had to toast them before a hot fire and soak them in vaseline before he could walk. This and the two following days were spent in getting the remaining skin and the skeletons into camp and dressing them. These five days were spent in securing two specimens, and as the three last days were in camp, Dyche was well rested and started on the sixth day for the mountains. While travelling down the mountain a goat made its appearance about half a mile away on a little flat-topped spur which extended out from the main ridge. The animal was feeding and could have been easily stalked from the main ridge, but the wind was unfavourable, and Dyche concluded to crawl down one cañon and up another, under cover of the crag on which the goat stood. Reaching the spot where the animal had been feeding, the naturalist could see nothing of it. An examination showed where a nanny and a kid had been feeding, and as they were just what Dyche wanted, he proceeded very carefully along the cliff. Hearing the sound of rolling stones ahead of him, he hurried forwards and saw an old she-goat with a kid on the other side of the cañon. With the crack of the rifle the old goat pitched forwards and rolled down the slope, while the kid hid behind some rocks. After waiting fully five minutes for the kid to come out, the hunter changed his position and the little fellow jumped out and started to run. The gun cracked again, and the kid went rolling down to join its mother. It made a sheer fall of fifty feet. The skull was crushed, the jaw broken in several places, the leg and shoulder broken, the skin torn, and worst of all, the horns were broken off and one lost. This almost ruined the specimen, but Dyche spent twenty minutes searching for the horn, and was fortunate enough to find it. The kid was measured and skinned, and then the search began for the old one. She had lodged on a projecting ledge, and the naturalist was compelled to work for some time to dislodge her. He was finally compelled to get a long pole and make a ladder by which he got down to the goat, and here he was compelled to tie himself to the rock to prevent a fall while he worked over the specimen. Dyche was so worn out that when he reached the foot of the mountain he cached the specimens, and with only a portion of the flesh of the kid he went on to camp.

He found, however, that the flesh, even of the kid, was hardly fit to eat, for it was so permeated with the musk of the glands that it was unpalatable. On this trip Dyche was fortunate enough to find a spring of pure water on the range, probably the only one on the mountain, and this was always made the objective point of all trips hereafter.

CHAPTER XIII.

A Peculiar Danger—Four Goats in Four Shots—A Rapidly Disappearing Tribe—Description and Habits—A Persistent Hunter.

EANTIME the mountain fires had been steadily increasing and advancing until the camp was threatened, and Dyche began to fear that all his labour would be in vain, for if the fire came down the mountain the skins and skeletons would be destroyed. It was with a feeling of great joy that the naturalist awoke next morning and saw the rain pouring down. It was the first of the season in that locality and came in such quantities that the fires were soon checked and then were drowned out entirely, with the exception of smouldering logs here and there. The air cleared, and the thin spiral columns of smoke ascending from various points over the mountain were the only evidences of the conflagration that had lately been raging within a few miles of the camp.

Dyche at once started after his specimens and brought them into camp, putting the skins in a pickle. It continued to rain, and soon the tops of the mountains took on a mantle of white; and as the rain increased, the snow-line crept lower and lower, until it reached the green of the timber. Now another dan-

Almost despairing.

ger menaced the hunters' camp. The peculiar formation of the mountain caused great masses of rock to split off, owing to the action of the water, and immense bowlders were continually rolling down into the cañon. The stillness of the mountain would suddenly be broken by an explosion resembling the rattle of musketry, and then a great mass of rocks would lean out from the face of the cliff and topple over, falling, rolling, and tumbling to the bottom and sometimes going as far as the river. These rock slides would crush everything before them, and immense trees and bowlders were torn from their foundations and hurled with the force of an avalanche to the bottom.

This was very dangerous, for if such a slide ever struck camp there would be nothing left to tell the tale. One incident convinced the naturalist that he could not get away too soon. It was after midnight when Dyche was aroused by a peculiar roaring and rumbling noise directly above their heads on the side of the mountain. The two men were out of their sleeping-bags in an instant, and a run of thirty yards placed them behind an immense tree which had previously been selected for just such an emergency. The tree was over five feet in diameter, but even then Dyche feared that it would not withstand the terrible force of the rocks. An immense landslide had broken loose half a mile above the camp and came roaring down with the speed of an express train. It passed a few feet to the south of the camp and expended its force among the bowlders and in the river. Sleep was over for that night, for they could not tell how soon another mass might come down, nor did they know until morning that their horses had escaped.

The continued rain and snow and the falling rocks convinced Dyche that he had better finish his hunt as soon as possible and get away. He had two fine skeletons and four skins, and the naturalist would have been satisfied with these alone, for he had as many as were possessed by all the museums of the world; but he felt that he ought to try once more to get specimens while he was so near the fountain-head. Next morning he went over the ground traversed during the first hunt. It had ceased raining and everything was covered with snow. The naturalist wandered along, feeling repaid for his trouble in the beauty of the mountain-tops after the snow-storm. But his meditations on nature were soon diverted. At his feet he saw a goat-track plain and fresh in the snow, and this he followed over the rocks until it seemed to be directed towards a bold crag which jutted out from the side of the mountain.

The tracks led directly to the jutting crag, and when that was reached Dyche was amazed to find that the animals had gone around the cliff on a ledge which a dog would have been unable to traverse. The little shelf was but a few inches wide. On one side was a sheer descent of hundreds of feet, while the perpendicular wall rose to the clouds on the other. The appearance of the place was such that it seemed possible only for an animal with wings to go around it, yet those goats had gone over the ledge as calmly **as if** it were their usual **route.** The naturalist found that he must retrace his steps and ascend to a higher ledge before he could work around the crag. Having done this, **he** was moving slowly along the side of the mountain, keeping good watch over the country, when he saw four goats on the opposite side of the cañon.

The goats were over half a mile away, and were evidently those which **be** had followed to the crag. They were not as watchful as the sheep which he had hunted, for they fed along with heads **down.** An hour's stalk took **the** hunter to **the spot** where the goats had last been seen, but **they** were gone. Fearful that they had taken the alarm, Dyche walked carefully along the ridge. Looking over into a little draw, he saw them all feeding in a grassy spot on the opposite side of a deep cañon. Another hour's hard work brought the naturalist near the game. The animals were now in plain sight, not over two hundred yards away. Skulking and crawling the hunter **reached a log, and** behind this **he lay and** watched the goats.

His **first** impulse was to shoot, for they were within easy range, but a moment's reflection showed him that they were feeding towards him. They moved slowly, feeding part of the time and then moving about aimlessly, but all the while coming nearer and nearer. At last they were within a hundred yards and in a comparatively clear place. Carefully examining his gun to see that it was in good order and that the sights were in place, Dyche prepared **for** the shot. He examined the ground and decided on the place to which he would jump when he fired the first shot. Training the gun on a low shoulder-spot on the first goat, he fired and then jumped to his feet. The other three animals were going at full speed towards the timber. A yearling ran along a log and received a shot as it turned its side to the hunter. It rolled from the log, and a shot was sent after the hindmost of the other two, which immediately sank down. The old nanny was over two hundred yards away by this time and was nearing the top of the ridge, but as the gun cracked she tottered and fell. Four goats now lay dead in a straight line up the hill, killed with four shots. Dyche could not restrain his enthusiasm, and bounded up the hill while the mountains echoed with his shouts.

An examination showed that four more fortunate shots could not have been made. The first went through the shoulder, the second through the liver, the third went high, striking in the hip and ranging forwards under the ear, while the fourth also went high, through the shoulder and ranged forwards. It was now half-past three, and by hard work the entire load of skins was taken to camp by half-past eight. After supper was over and his shins and knees were well rubbed with vaseline, the naturalist sat up until midnight dressing the skins. Next day, while going for the bones, a fine mule buck was killed, which supplied the camp with meat. Three or four days were spent in getting the specimens to camp and in caring for the skins and bones properly, and then another hunt was taken over the grounds, which were now becoming familiar to the hunter.

An early start placed Dyche on the top of the mountain by nine o'clock, and a magnificent spectacle was presented to his view. The heavy snow extended down in irregular tongues of white to the velvety green of the trees below. Lower down the frost-touched quaking asps gave a glorious colouring to the scene with their leaves of green and gold, just in front of the sombre background of burnt forest on the opposite slope of the mountain. The clouds had passed away and the sun shone with unwonted brilliancy, while the pure and invigourating air gave life to everything. This was the home of the Rocky Mountain goat. Here he lives, away from all other animated nature, and here he should roam for ages to come. He has chosen for his home the land where rugged crags and peaks stand as an almost insurmountable barrier against other animals and even man is kept aloof. The sparse grass and such shoots as he feeds on are out of reach of other ruminant animals, while the bleak and dreary mountains offer no enticement to the carnivora. He is harmless and is almost worthless as far as food is concerned; his skin is so tender that it is valueless for commercial purposes. All these facts being taken into consideration, the natural supposition would be that the Rocky Mountain goat would increase and become common on all the high and bleak mountains. But such does not seem to be the case.

The goat, at one time, is said to have roamed the high mountain-tops of the Rockies and Cascades from Northern Colorado to Alaska, but now he is found only in the most inaccessible places of the far northern mountains. This range never exceeded four hundred miles in width, and it is doubtful if the southern limit ever reached Northern Colorado. Hunters tell of the goats which they have seen in Northern Colorado, but it is probable that they have mistaken bands of female sheep for goats, which they greatly resemble at certain seasons of the year. It is certain that the animal abounds in parts of British Columbia, but its capture is attended with so much hardship that it is rarely sought for except by those naturalists who are something more than room-workers. The goat will be found for years by those who are hardy enough to search for him, but there will be few killed.

It was while such thoughts as these were passing through the mind of Dyche that he saw a goat walking on a ridge about half a mile away. The animal went to the point of a stony crag, from which it viewed the country for ten or fifteen minutes. It was so far away and the country was so rough between him and the animal that the naturalist was in some doubt as to the best mode of procedure. The goat solved the question by descending into one of the stony cañons, and as soon as it disappeared from view Dyche began to stalk it. The goat came out on another crag and looked around. It appeared uneasy, and the hunter feared that it had winded him. From this crag the goat descended to a still lower one, and thence on down, until at last it appeared at a distance of about three hundred yards.

Turning to the right the goat began ascending a ridge, and calculating that its next appearance would be on a certain stony crag, Dyche **ran towards** this at full speed, dropping down behind a clump of spruce trees just as he thought the animal would come out. He waited about two minutes, and the goat walked **out** on **a** ledge opposite to the naturalist. The goat made a beautiful picture with **his** almost white body outlined against the gray granite of the mountain, not over **one** hundred and twenty-five yards away. He was across a cañon, which was always an objection with Dyche; but a steady **rest** across a rock gave a good aim, **and the ball sped** to the white spot on the animal's breast.

The goat staggered forwards and tumbled over the cliff, rolling to the bottom **of** the precipice, out of sight. Hastening down, the naturalist found a fine specimen. It was not so large as the first one killed, **but it** was in much better condition as to hair and colour. It was the whitest male that he had killed. The colour of the goats is usually described as white, but the hair has a yellowish tinge. Many of them have a very dirty appearance, for the long hair is filled with spruce needles, **cone** scales, dirt, and **even little** stones, which they accumulate as they rub in the dirt. Young kids have a streak of brown hair running down the back and brown hairs scattered all over the body. Yearlings show some of these brown hairs, but as the animal grows older they almost entirely disappear. A thick coat **of under** hair, **or** wool, is always present. **In** August the skins show a splendid short growth of this, as then the long hair is thin, most of it having been shed. **The** long, coarse hair of the goat gives to it the **ap-**

pearance of a small albino buffalo, except that the buffalo has long hair on the fore-legs and shoulders only, while that of the goat is long all over the body with the exception of the lower parts of the legs and upper part of the face.

The horns of this specimen were finely shaped and without blemish. All goats have horns, those of the old males reaching six inches in circumference, while those of the females are but slightly smaller. The horns stand well up, curving backwards and slightly outwards. Behind them are the musk-glands, which on the old males stand up two inches and are so strongly impregnated with musk that they taint the whole body, making the flesh unfit for food. Near the horn these glands are bare and scaly, but away from this part they are covered with hair. When cut open they show a pink colour and give out a strong musky odour.

The goat's nose is covered with short hair to the tip, with the exception of a narrow median streak between the nostrils. The eyes are straw colour in the living animal, but turn to a dark chestnut a short time after death. Under the throat the coarse hair hangs in a beard similar to that of the common goat. The hair on the legs is long, as well as that on the hump between the shoulders. This hump is one of the peculiarities which has long puzzled naturalists who have not taken the trouble to see the animal in his native haunts. It is simply a lengthening of the spinous processes of the dorsal vertebræ. The tail is very short and the ears small and pointed. The latter are evidently of little use to the animal so far as hearing is concerned, for in every goat that was killed on this trip the ears were so filled with ticks that the insects formed a compact plug which it was necessary to force out with a hard stick when cleaning the skins. The feet **are** almost square and the outside rim of the hoof is sharp, making a knife-edge which enables the animal to grasp any crack or opening and thus obtain a firm foot-hold.

The most common error of naturalists who have **never seen either a** living or dead Rocky Mountain goat is in regard to the position of the head. **In nearly** all the cuts of the animal in books or magazines the head is raised **above the** line of the shoulder. The anatomical structure of the animal clearly precludes this. The neck is short and set so low down that the head cannot be raised above a line even with the top **of** the shoulder. Taken as a whole, the proportions of the goat are similar to those of an American bison, or buffalo, on a small scale.

The specimen was skinned and the skeleton **laid** bare, and then **the** naturalist attempted to carry **the** whole load of skin and bones to camp. The load was a heavy, awkward bundle, but he made his way along, and by four o'clock in the afternoon reached a point on the mountain above the camp. He was tempted to leave his burden here and return for it next day, but the thought that it meant a whole day lost caused him to stagger on, and he finally reached camp long after dark. He found that there was not much in a load of bones and skins to arouse any **but a** scientific enthusiasm, and he spent several hours after supper in dressing the skins and rubbing his legs with vaseline. Next day was Saturday, and he remained in camp looking after the specimens and fighting magpies.

Provisions were now very short, and the place was becoming more dangerous with each succeeding day. The guide was eager to hurry away, but Dyche could not leave until he secured more specimens. Several days were spent on the mountains without success, and the naturalist was almost ready to listen to the advice of his companion. One morning after several hours' fruitless hunting Dyche was crawling carefully along a ridge, when he saw a goat lying down on the point of a ledge. When within seventy-five yards a shot was sent after the animal, which staggered to its feet and then pitched forwards and rolled over the rocks down the mountain. The hunter hurried to the spot, but could find no trace of his prize. After searching for some time he at last descried the dead animal lodged on a little shelf about a hundred yards below the spot where he had shot it. For two hours he tried in vain to get to the animal, but he was finally compelled to give it up. It was a great loss, as it was a fine young male and was needed to complete the series.

About the middle of the afternoon another goat was seen. He was standing on a crag above the hunter, looking down. Like all goats, it did not appear to apprehend danger from above. A shot through the shoulders sent the animal to earth, and it was after dark when Dyche reached camp with the skin and skeleton.

Next day while hunting along the mountain a bear-

trail was seen and Dyche started to follow it. The
bear had picked out a good trail and travelling was
easy. While hurrying along the hunter caught sight
of a goat standing on a crag above him. It was apparently looking directly at him, and Dyche feared
that he had been seen. Working his way out of
sight he tried to get above the goat. He climbed to
within a few hundred yards of the top of the mountain and there found that farther progress was effectually barred. Retracing his steps, he finally found
a way up and reached the top. With feet and legs
soaking wet from the damp snow, he found himself on the crag where the goat had been standing, but the animal was nowhere in sight. The
naturalist sat down to rest and eat his lunch, and
while doing this he noticed the goat standing on
the point of a crag below him. Working carefully, he began to descend. A point about two
hundred yards above the animal was reached, and
then a careful stalk was begun to get closer, for the
snow was falling too heavily to risk a shot at that
distance. Crawling down to within one hundred
and twenty-five yards, a shot was directed down at
a point between the goat's shoulders. It did not
move, and Dyche feared that his shot had been a
clean miss. Hastily slipping another cartridge in
the barrel, he was just about to fire when he
noticed a red stream running down the shoulder.
The animal began staggering, and fearing that
it would pitch forwards over the cliff, the hunter
sent another shot at it, trying to break its back, so
that it would sink in its tracks. This effort, how-

ever, was a failure, for the goat pitched heavily forwards and rolled down the mountain. Making his way down, the naturalist found that he had killed a large dry ewe, one of the finest specimens that he had obtained. It was in perfect condition and was fully as large as an ordinary male.

Two more days were spent in the mountains in unsuccessful hunting. The horses were running down in flesh so rapidly and provisions were so low that the necessity of leaving became at last deeply impressed on the mind of the naturalist, and arrangements were made to start. But Dyche was determined to have one more hunt. He started before daylight for the mountains, in a heavy snow-storm. Finding a bear-trail, he followed this some distance, when it was crossed by a goat-track. Experience had long since taught him the futility of following a bear-track, so he turned his attention to that of the goat. He followed it for about half a mile, when the animal was seen going down a ridge. The goat saw the hunter at about the same time and started at full speed over the rocks. Dyche followed the trail in the snow, hiding as much as possible, and at last saw the animal standing on a projecting point, looking back. After long and circuitous stalking it was found that the goat had hurried on up the mountain. The track was again followed until it entered a piece of timber. Here another circuit was made and tracks of bears, wolverines, and lynxes were seen, but the goat had not left the woods. Dyche began circling around the mountain, watching carefully. While creeping along in the timber the hunter caught a glimpse of his quarry just as it was disappearing among the trees. A quick shot was sent after it and then it was lost to view. Hastening to the spot, he found the tracks and soon saw spots of blood. Following these about three hundred yards he came upon the animal, lying down. Its expression was anything but mild, and to avoid accidents the hunter sent another ball into it, killing it instantly. It was a fine two-year-old male and was just what was needed to complete the collection. It was now raining hard, and after two hours spent in taking notes and measurements and skinning the animal, Dyche threw the skin across his shoulders as a protection from the rain and started for camp. It was almost dark and the distance was long, but the hunters were to move the next day and it would not do to leave the specimen on the mountain. He staggered along and finally reached camp with his precious load. His knees were in a fearful condition, but he rubbed them until they felt better, and then spent several hours preparing his specimens for packing.

The horses had had a long rest, but were in bad condition from lack of proper food, and it was a hard trip over the mountains to Thorp's ranch, which was finally reached with as fine a lot of specimens as was ever carried out of the country.

CHAPTER XIV.

On Kettle River—Two Model Camp Companions—A Royal Deer—Eating Beaver Tails—A Tramp over the Mountains—Wolves about the Cabin—Varieties of Deer.

THREE days were spent at Thorp's ranch, recuperating from the hardships of the goat-hunt. Specimens were stored and packed, and then arrangements were made with Maurice Farrell and George McLaughlin for a second trip up Kettle River. These men were old mountaineers and knew every foot of the country for many miles around. They were sure they could take Dyche to where he could find caribou. They had spent the previous winter at the cabin mentioned in a former chapter, which they had built, and they had brought out with them in the spring over a thousand dollars' worth of furs. They had frequently seen caribou near the cabin, and Indians had told them that they were abundant on the mountains above.

These stories made Dyche believe that he could accomplish the entire object of his trip to this country. The start was made with a string of a dozen pack-horses. McLaughlin's readiness in finding a trail through the mountains and forests had earned for him the title of Pathfinder, while Farrell was considered the best campman in the whole country,

Off for a long tramp.

his especial forte being the preparation of a meal. He demonstrated on this trip that he fully understood the art of camp-cooking. Pancakes were his specialty, and these were cooked in a style that would put many famous chefs to shame. With a frying-pan in each hand, he would keep them turning and twisting with a simple movement of the wrist until the cake would turn over with a precision that caused each part to be perfectly cooked. Nor was anything else neglected while this was being done, for the trout or venison steak was cooked to a turn, and the coffee and biscuits were never known to be amiss.

The first camp was made at Wannacut Lake, and while the Pathfinder and the naturalist unpacked the horses and arranged the camp, Farrell prepared supper, the first regular meal of the expedition. Fried ham, boiled potatoes, fruit, pancakes, and coffee were soon ready, and the signal which was always used on this trip was given:

"Square yourselves for action, boys."

They "squared themselves," and the principal comment was that they wanted more. The alkali water of the lake so embittered the coffee and pancakes that it was hard to swallow them, but they managed to make a good meal. The village of Wannacut was about a mile from camp, and after supper Farrell went there with a bucket for water with which to cook breakfast.

The horses had a good supply of water and grass at Rock Creek, and next day at two o'clock in the afternoon Kettle River was reached. Here Farrell and

Dyche left the outfit and went to a Chinese settlement and purchased eighty pounds of "spuds," or potatoes. These were placed in two bags which they carried in front of them on the saddles, and then it took all the rest of the day to catch up with the Pathfinder, who had gone ahead with the train. On the way "Kettle River Armstrong" was met, a ranchman who devoted his attention to raising horses on a ranch on the middle fork of the river. Armstrong said that caribou were thick about his ranch. He had seen where four had passed by a few days before. He tried to induce the hunters to go to his ranch, but they decided to follow their original plans. Their route lay into the wilderness described in a previous chapter, and at the forks of the river McLaughlin had made camp and was awaiting them.

A heavy rain fell next day as they prepared to start, but no one seemed to think any the worse of the weather, and jokes were flying thick and fast as the trail was taken. A deer became too curious and his venison swelled the larder. Camp was made that evening in the forest, and while Dyche and McLaughlin were attending to the horses, Farrell was preparing a supper fit for the gods. Farrell compelled the others to strictly observe one thing, and that was that they should not come "putterin' around the fire." They willingly acceded to this demand, and only when the cry of "Square yourselves for action" was given did they venture near.

Another day of travel took them to the cabin, and here the two men were at home. Sitting around the fire they related enough of their experiences to fill scores of books of adventure and hunting. They told how they had secured thousands of dollars' worth of furs with their traps and guns, how fifty beaver and numerous wolverines, lynxes, otters, fishers, martens, wolves, and other animals which came their way had been captured. They told of the gaunt, black timber-wolves which roamed in great bands through the trackless forests and destroyed many deer; how it was almost impossible to trap or poison them, for they were too shrewd to eat poisoned meat or go near a trap; how a pack of the big hungry brutes would kill and eat a deer and leave no vestige of it except a few scattered hairs on the snow. The very bones would be crunched and swallowed by the ravenous beasts.

McLaughlin showed where he had stood and seen a fine buck dash from the forest and run directly towards him, its tongue hanging out as it panted from the great exertion. The presence of man did not frighten it, for it was fleeing from a more deadly enemy. As it passed on a pack of howling wolves burst from the woods on its trail, and it was not until three of them fell before the balls from the trapper's Winchester that the ferocious brutes turned back into the depths of the forest.

McLaughlin told of that mysterious animal, the wolverine, which the Indians have so aptly named the "mountain devil;" how the trap must be fastened to swinging poles or the animal will carry it away. Even when the trap is chained to a pole the wolverine sometimes climbs the chain and gnaws the pole through, carrying off the whole load. One got away with a small trap, and two weeks later his fresh trail was seen in the snow and the trap was still dragging behind him. Another carried off a larger trap a few days later, and then the Pathfinder concluded it was time to do something to prevent the wolverines from "setting a line of traps," and he took a pair of blankets and two days' provisions and started on the trail. He travelled fifty or sixty miles, but failed to catch the robber.

Two days were spent at the old cabin, and Dyche was well pleased with the energy displayed by his new companions. McLaughlin went four or five miles up the river hunting for a new camping-place, and seeing a fine mule buck standing in a suitable place he killed it, "just to mark the spot." It was snowing heavily next morning when he and Dyche started out to see what was in the country. The trappers had told the naturalist of a big buck which they had seen here several times, but the animal was so wary that they had never been able to get within gunshot of it. Their description made him a veritable giant. The dense woods had great fascination for the naturalist, for here he felt that he might find new animals at almost any moment. In three hours he reached the top of the mountain and there found numerous tracks of bears, wolverines, fishers, and deer. Many deer were to be seen standing about in the woods, and now and then they would move away with long, graceful strides for a few hundred yards, and then stand and gaze at the intruder with a "wonder what you are?" stare.

As the hunter approached the edge of the woods he

saw about a dozen deer standing on a ridge. When they moved away Dyche caught sight of an enormous buck which led the band. He stood a full head and shoulders above the rest, and at once the naturalist thought of the big buck of which McLaughlin had spoken. Attempting a stalk, he soon found that the animals were too wary. He did not like to be outwitted, and started back on his trail until he was well out of sight, when he began a wide detour in order to get ahead of the band. After travelling over a mile he reached a spot where he thought he was ahead of the deer. While moving quietly through the woods he caught a glimpse of a band of deer in front of him. As they were looking directly at him he stopped. Just then the big buck came stepping proudly from the timber, gazing in the direction of the hunter. As he stood with head erect he made a magnificent picture, but the naturalist lost no time looking at it. He sent a bullet through the animal's breast and it dropped at the crack of the gun.

This was a fine specimen, the largest deer that Dyche had ever seen. There was not an ounce of fat on him. If he had been in the condition that he undoubtedly was earlier in the year he would have appeared as large as an elk. The measurements were taken carefully and they were something wonderful for a deer. His standing height was forty-four inches from the top of his back to the flat of his hoof. The circumference of the body behind the forelegs was fifty-one inches; that of the abdomen was fifty-five inches. The skin, skull, and leg bones weighed forty pounds, and the naturalist carried the load four or five miles to camp, only to receive the severe censure of Farrell for making a pack-horse of himself when there were so many "cayuses" standing around.

McLaughlin was out three days, and when he returned he reported a few signs of caribou. One or two of them had been feeding on the bunches of black moss which hung from the trees at a place about ten miles from camp. The trail was taken in that direction, and after a hard day's travel a little meadow was reached during a driving snow-storm. Camp was made under difficulties, and it was some time before a big log fire made the hunters comfortable. With a bright fire, hot supper, and many good stories, they were soon laughing and joking as if there was no such thing as a snow-storm. As the night deepened a pack of timber-wolves made their appearance and serenaded the new-comers. In the deep stillness of the forest these weird howls brought many tales of danger to the minds of the campers, but they were not of a mould to be scared at snarling wolves and all were soon fast asleep.

Several days were now spent in exploring the country, which was found to be an immense jungle, and then Dyche and Farrell made an effort to reach the mountains on the east. From this place could be seen a great wilderness of trees, but off to the northeast appeared the shimmering water of several lakes. The bald mountain to which Dyche had travelled on foot several weeks before reared its snow-covered head above all the others. After long consultation that night Farrell and McLaughlin determined to cross the range and go down the middle fork of the river to a point about fifty miles above Armstrong's ranch. The party at once set about to find a way out of the cañon, and after a whole day's chopping a path was cleared. Meanwhile Farrell had set a lot of beaver-traps and caught two fine animals. The tails and hams, cooked with beans and pork stew, made a most palatable dish for the hungry hunters. The tail was considered a great luxury.

The trip over the range next day was enlivened by Dyche's horse, Chief, which suddenly ran away and bucked off the pack of cooking utensils. The outfit was scattered over the mountain and Farrell took the affair as an especial insult to himself. After indulging in very strong mountain language about the horse, he offered to present the naturalist with a good one if he would give the runaway a ball from his Winchester. Hard work was the rule next day, and evening found them still on the side of the mountain. McLaughlin's great exertion with the axe and his abstinence from food during the day now had a bad effect, and he was suddenly attacked by cramps and he rolled on the ground groaning in pain. Dyche and Farrell ministered to his trouble as best they could, and by dint of rubbing at last succeeded in relieving him so that he felt able to travel. Before grass and water was reached, however, the cramps returned with increased severity and they were compelled to camp in the big woods. A fire was made and the sick man rolled up in a bundle of blankets and warmed, but he suffered intensely, and it was only after long and vigorous rubbing that he became easier. Farrell found that they were within two hundred yards of the river and water was soon secured for camp use.

Next morning the Pathfinder crawled out and went to work despite the advice of his companions. They were now in a magnificent forest. Trees as straight as the masts of a ship rose from one hundred to two hundred feet in the air, with tops so matted together that it was almost impossible to see daylight through them. It was decided to go up the river, but six miles' travelling in this direction convinced them that they could do nothing there and they retraced their steps. There was little sign of large game. A peculiar willowy bush grew from the ground to a height of three or four feet and then the tops bent over and again took root, making an almost impassable underbrush.

Armstrong's horse ranch was the next objective point. On the second evening Farrell went out to get some fresh meat. Soon a shot was heard.

"That means fresh meat for supper," said McLaughlin.

As he spoke another and then a third shot rang out.

"I'm not so sure of it now," was the comment.

Then followed several shots in quick succession.

"That much shooting means nothing, or maybe a fawn," was the remark. That he well understood his partner was shown when Farrell came into camp a short time after carrying a fawn across his shoulders. The joke was on Farrell, but next morning he went out and found two bucks that he had killed the night before.

Ready to go home.

Next day Armstrong's ranch **was reached** in a blinding snow-storm. **An** abundance of grass gave the horses good food and they were turned loose to shift **for** themselves. The first day was spent in building a tent made of poles and spruce boughs, called in that country a rancheree. A few days were now spent in prospecting in every direction. Many **deer** were seen daily near the rancheree, but **there were** no signs of **caribou.** Each night Armstrong came over **from** his ranch, about a hundred yards away, and the mountaineers told stories of their life while Dyche cleaned and prepared his specimens.

The Pathfinder now began **to** show great signs **of** uneasiness. He had been told that there was plenty of big game in these woods, including caribou, but he had so far failed to find anything but a number of **deer,** which were too common to shoot. He felt that something must be done, and one evening, after sitting brooding awhile over the fire, he stretched himself, yawned, and then said:

"I'm going to **find caribou or be satisfied there's** none in the country. **To-morrow I'll** leave you for that big mountain over there, and I'll see what's become of all the animals of this neck of the woods."

"How will you go?" asked Dyche, who at once became interested in the new turn which things were taking.

"Walk," was the laconic answer.

"I'm with you." So it was settled that a long tramp of thirty or forty miles should be taken over the snow-covered mountains, and preparations were **at once** begun. Two pair of blankets **with** provisions sufficient to make a load of about forty pounds were put into a pack for each, and the start was made early next morning. Finally the mountain-side was reached and the long and steady pull began. A small frying-pan and a coffee-pot hung to the packs, and at noon a halt was made to get a cup of tea and something to eat. At half-past three the Pathfinder called out:

"Here's the place for us. We'll camp."

Three huge dead trees had fallen across each other beside a broken stump about twenty feet high. The ground was cleared of snow and fires were started to dry the earth. Soon the hunters were enjoying a supper of venison steaks, biscuit, and coffee. At eight o'clock both were asleep, and by four next morning they were again on their way up the mountain. Two more days of travelling through the dense underbrush, with snow eighteen inches deep, took them to timber-line, and the bald top of the mountain rose before them. The snow was badly drifted and it was very cold away from shelter. A circuit to the north was taken towards Kettle River, but no signs of caribou were to be seen.

Their curiosity was fully satisfied, and, as they were fifty miles from the home camp, they began to plan their return. They readily saw that even if they found game there would be no way of getting it out of the country, for the horses could never be brought up the mountain. It was now the middle of November, and another foot or two of snow was likely to fall any night.

Next day a difficult piece of country was entered,

and the travelling over fallen timber and rocks and through deep gulches **was very** hard. **Two** grouse furnished **supper** for the hunters and this was all the meat they had. They were now well down the mountain, and just as they were preparing to go into camp they saw an immense bear-track. The animal had passed along but a short time before. The track was fully nine inches wide in the soft snow. As he walked he had **swung** his claws out to one side, and the marks left in the snow were enormous. **This bear was just** what Dyche was needing, **and the en**thusiasm **of** both the hunters was aroused.

"That old fellow will **not** go far in this weather. He's **just out for an** evening's walk. I'll bet we'll find him within two miles of camp in the morning," said the Pathfinder.

"**Well, I'm** going to follow him to the jumping-off place or get him. He's just the one I need for my collection," replied Dyche.

"Oh, we'll get him sure enough. Don't you worry," was McLaughlin's encouraging answer.

"**I'll tell you** what I'll do," **said the** naturalist. "I'll give you a clean fifty-dollar bill extra if we get that bear."

While McLaughlin was eager enough to go after bear, or anything else that Dyche wanted, without the offer of extra pay, this inducement filled him with a **desire to** slay the bear at once. The two hunters spent the night dreaming of big bears, but their hopes were sadly dashed when they saw a heavy fall of snow that had come down during the night, completely obliterating every vestige of the bear-track. They were almost inconsolable, and after taking a big circuit in order to see if the old fellow had been moving that morning, they were compelled to return and reluctantly take up the trail for the rancheree. They reached the river about five miles above the home camp, and as it was too wide for a foot-log they searched for a place for a ford. Reaching a point where the river widened to about two hundred yards, McLaughlin said it was shallow enough to wade and they entered the water.

As the Pathfinder plunged into the stream he gave an involuntary "whoop!" for it was like ice. The cold almost paralysed the muscles of their legs. McLaughlin went ahead and Dyche followed close behind, pounding and pushing the ice out of the way, while their teeth chattered. Just as he reached the middle of the stream McLaughlin suddenly stopped and began making such strange signs that Dyche's heart almost stood still with fear, for he thought that the Pathfinder had been attacked with another case of cramps. If this was the case then it meant death, for there was no possible way of relieving him there. These fears, happily, were groundless. Mac's queer actions were caused by an effort to straighten his pack, which had slipped.

The shore was finally reached, and then it took them half an hour to rub vitality back into their half-frozen legs. When this was done they hastened down the stream to the home camp five miles away. As they hurried along they saw a buck standing in the edge of the woods and both fired at him, bringing him to the ground. He was almost as large as the big fellow which the naturalist had killed when he first came into the country.

The **rancheree** was reached just at dusk, **and** Farrell soon had a smoking supper ready, which drove from their minds the hardships of the trip. A week was now spent in gathering large and small animals, Dyche dressing and preparing the skins and skeletons supplied by the two mountaineers. Twenty-one choice specimens of deer were added to the collection. Armstrong took all the extra meat, storing it **away** for **use in** the spring, when venison **would be** in bad condition.

The woods abounded in the large timber-wolves, which roamed about at night and remained well hidden during the day. Every night they could be heard howling just outside the tent, and they chased Armstrong's **dogs** to his very door and attempted to reach them under the shelter of the sheds. Poisoned meat and traps had no terror for them, as they instinctively shunned them. One evening Armstrong heard an uproar outside of his door and opened it **just** in time to shoot a large black wolf which was attempting to kill his dog within ten feet of the house. Thus the week **was** passed, and at the close of the time allotted for the stay the naturalist began to get his specimens in order for the trip down the river. The last evening Armstrong, Farrell, and McLaughlin each brought in a deer. They were so peculiarly marked that a discussion of deer was started at once.

Armstrong's deer looked much like a common or Virginia deer, but the tail was black on the outside and white underneath. It was thought at first that this was a true Columbia black-tail deer, but McLaughlin, who had killed many of them, was not sure of it.

"Why do you call these mountain deer 'mule deer?'" asked Farrell. "They are called 'black-tails' all through this country."

"Yes, they are usually called 'black-tails' by the hunters up in this country, but they are not the true black-tail deer. The Columbia or true black-tail deer live on the Pacific side of the mountains, ranging from California to British Columbia. Their range is thus restricted, and only stragglers are ever seen on this slope."

"I've killed lots of them on the coast side," said McLaughlin. "They are not so large as these mountain or mule deer. Their legs are shorter and their ears are not so large."

"There's where the distinction comes in," said the naturalist. "They are not as large as the common Virginia deer."

"Well, I'd like to know how a fellow's going to know them when he sees them," said Farrell. "Some mule deer, as you call them, are small, with short legs."

"They can always be distinguished, if you know how to look at them," was Dyche's answer. "The tail of the true black-tail deer is black or tawny black on the outside and sides, and there is a streak of white underneath. The tail is round and full-haired, much like that of the Virginia deer. That of the latter is flat, however, and much longer than the other. The ears of the Columbia deer are about half-way between those of the Virginia deer and the mule deer in size. The antlers are almost exactly like those of the mule deer."

"Is there any distinction between the horns of the different species of deer?" asked Mac.

"There is a greater difference here than anywhere else except in the glands of the legs. The horns of the Virginia deer rise from the head, swing back and up and around to the front of the head as one beam. The points, or branches, all grow up and out of this main beam. The first or brow point is usually rather long. The horns of a mule deer come up in the same general way, but spread more and the branches do not stand up on the main beam. The brow point grows in the same way as in the common deer, but is usually shorter. In the mule deer, you see, the main beam divides. Here you have it in this one," said Dyche, pointing to a set of antlers near. "Here it branches into two equal parts and these branches again divide equally. This is the distinction and is constant. A brow point and four branches, usually of the same size, give the general plan of the antlers of the mule deer. There may be several other branches and snags growing from the horns, but the general plan can usually be made out without trouble if you look for it."

"Well, that clears me up," said the Pathfinder. "As I now understand you, the horns of a mule deer, or true black-tailed deer, branch about the same way and there are four main prongs besides the brow point. If the deer has such horns and has a round tail about as large as a mule-deer tail, black or blackish on the outside, a white streak underneath, a full-haired tail, not bare underneath like a mule deer, and round, not flat like a Virginia deer, then the animal is a true Columbia black-tail deer."

"That's the idea," said Dyche, "but you should not forget to take in the ears and leg glands. The ears of the common deer are about half as long as its tail, the ears of the mule deer are nearly as long as its tail, while the ears of the Columbia deer are of intermediate size, being a little over half as long as the tail. Then there are the leg glands, which are always a distinctive mark. The metatarsal gland is situated on the outside of the hind leg between the hock joint and foot. The gland is easily found from the extra length of the hair which grows about it and forms a tuft. The glandular structure can easily be seen by separating the hair which grows on either side of it. This gland in a mule deer is from six to eight inches long, extending down from just below the hock joint; in the true black-tail deer it is from three to four inches long; in the common Virginia deer it is from one to two inches long."

"Well, I'd just like to know what kind of a deer this is that Armstrong killed," said Farrell. "Its tail is black on the outside."

"You see the horns are those of the common Virginia deer," said the naturalist. "There is one main beam, with long five-inch brow points. All the other points grow up from the main beam. The tail is, as you say, black on the outside, but it is long— twice as long as the ears. Horns, ears, and tail go to show that this is a Virginia deer. It is not an uncommon thing for the Virginia deer of this northwestern country to have the outside of the tail black. Deer in the northern part of North America, as a rule, are much larger than those of the southern climes. Bucks up here frequently reach a weight of two hundred and fifty pounds, while one weighing one hundred and fifty down there is considered a big fellow."

"What's the difference between the long-tailed deer, the white-tailed deer, the Virginia deer, and the common deer?" asked Maurice.

"A difference in name only," replied Dyche. "The species ranges all over the United States and up in the mountains of British Columbia. Some writers have gone so far as to describe geographical varieties of this deer. The mule deer is peculiar to the western half of the United States. Its natural home is in the mountains, but it was formerly found along the wooded streams as far east as Kansas and Minnesota."

"You fellows would talk a pine stump deaf," interjected Armstrong. "What's the difference? A deer's a deer. The meat is all the same. I've been eatin' it for twenty-five years and oughter know something about it by this time."

"You'd better go and cut some wood for your *clutchman* [squaw]. As I came by I saw her chopping and splitting that old log in front of the house," replied Maurice.

"Yes, I was out huntin' all afternoon and she didn't chop any wood," said the ranchman. "She can chop as well as I can but waits for me to do it. When I got my supper I made a sneak and came over here and let her do the choppin'."

"Did you ever see a spike-horn buck?" asked McLaughlin.

"Yes; I killed a Virginia deer in New Mexico with his first horns. They were about six inches long and were single. The first horns of a deer are usually forked. Spike-horns are very rare."

"Do does ever have horns?" asked Farrell.

McLaughlin answered this by saying that he had seen one doe, in his twenty-five years' experience, which had horns. Dyche told of a skull with small horns that he had seen in Denver which had been represented to be that of a doe. The eyes of the men began to grow heavy now, and Armstrong went home, and the others were soon wrapped in their blankets listening to the music of the wolves and the wind in the pines.

The horses were rounded up with great difficulty next morning, for the animals had had a long rest with plenty of food and were full of life. Three days' hard travelling carried the party down the valley and over the range to Loomis' ranch, where Dyche found twelve letters. The last freight wagons of the season were on the point of starting for the railroad, two hundred miles away, and the naturalist made arrangements to have his specimens taken along. Maurice and McLaughlin helped him to the last minute, and the parting from them was like the separation of old friends. The acquaintance had been but of a month and a half's duration, but the camp-life had shown the strong qualities of both men. The management of the camp was second nature to them, and the trackless wilds of the great forest were as plain and open to them as are the streets and alleys to the dwellers in the cities. They loved the forest and its associations. To them the singing of the pines was sweet music. Trees and rocks were their companions. They were true sons of Nature and in touch with her ever-changing mood. Dyche was of the same mould; the parting was not an easy one.

CHAPTER XV.

At the Lake of the Woods—After Moose—A Plague of Mosquitoes—Dark Swamps and Deep Rivers—Compensations.

THE work of the winter and early spring in the workshop had put the specimens of the British Columbia trip in good shape for preservation. Several fine groups were mounted and placed on exhibition in the university museum. But as the season advanced the warm months brought to the mind of the naturalist a desire for more work in the woods. Dyche wanted a group of moose to complete his collection, and he laid his plans before the board of regents. When asked by them where he would get his moose he acknowledged that he was yet uncertain, but one thing was sure, and that was that moose were not to be found on the campus of the university. The regents looked with favour on Dyche's plans and told him to go ahead, and he at once began corresponding with people from Maine to Alaska. Moose could be heard of in different parts of the country along the northern border of the United States, but after carefully considering the claims of each place the naturalist decided that the country around the Lake of the Woods would offer the best opportunities.

The monarch of the woods.

The first of July found Dyche at Warren, Minnesota, where he met a hunter and Indian trader with whom he had been corresponding. This man, Brown, was not only a hunter, but also a naturalist and a close observer, and he was therefore a most valuable companion. While they were preparing for the trip into the swamps Brown told Dyche much about the mosquitoes, and these stories were so extravagant that the naturalist was disposed to look upon them as largely fiction, but he afterwards found, to his sorrow, that no stretch of the imagination could equal the reality in this case.

A few days were spent at Warren in buying an outfit of horses and a light wagon. Just as they were ready to start for the swamps, Loughridge, a rancher who lived at the end of the road towards Thief Lake, came to their camp. He was an old friend of Brown's and he asked the hunters to drive a team of mules which he was sending home. He wished the two men to care for the animals. They were willing, and he insisted that they should make his ranch their headquarters while they were in the country. Loughridge gave them explicit directions about the road and suggested that they stop the first night at the ranch of a Norwegian named Goshens, who lived about a day's travel from Warren.

As evening approached they began to look for Goshens' place and inquired of the Norwegians whom they met, but found that the English language was an unknown quantity in that region. At last they met a bright-looking young fellow and asked for Goshens. A stupid look rested on his face for an instant; then he replied with a smile, "Ja, Ga-shens."

He pointed in a most indefinite manner towards several houses which were in sight about a mile from the road, but as they were in different directions his information was useless. Some time was lost in trying to discover the fellow's meaning, and at last the party decided on a house and went to it. On asking if Goshens lived there they received an affirmative answer, but were surprised to find that no one about the place talked English, although Loughridge had said that Goshens could give them information. Arrangements were made, however, for accommodations.

Mosquitoes had been plentiful for hours before they reached the ranch, but now the air was literally alive with them. The horses were fighting, kicking, and rolling in their efforts to get rid of the pests, and when Dyche went to put blankets on them he was driven back into the house by the swarms of bloodthirsty insects and was compelled to cover his head with a thick cloth and his hands with gloves before he could get near the horses. Then followed a wretched night, the remembrance of which will ever remain vivid in Dyche's mind. It was a constant struggle for fresh air against swarming hordes of the most villanous mosquitoes that ever sung. The air was perfectly black with them, while their noise was almost like the wind in the pines. To make matters worse, red sand-fleas got into the sleeping-bags and did an active business all night long.

The second night was passed at the ranch of another

Norwegian. This ranch, or "camp" as it is called, was in a grove bordering a great marsh some seven or eight miles wide. These marshes are covered with grass and in the dry seasons haymakers are frequently seen at work there. The ground is virtually made of dried grass and ligneous deposits, which at times are burned away by fires that rage across the country, eating great holes in the ground and leaving a most uneven surface. Lakes and pools of various sizes in these marshes afford breeding-places for countless thousands of water-fowl, while over all, under all, and through all are the mosquitoes.

The third day's travel took the hunters to Loughridge's ranch, where they were in the heart of the mosquito country. Large smudges of old damp logs and hay were kept constantly burning to give the horses some relief. So terrible were the attacks of the insects that the poor animals would not leave the smoke to get necessary food. Next day the hunters started for Thief Lake, which was four miles away. The whole country was a jungle of brush, logs, and pea-vines, and evening found them still on the road with the lake half a mile away. Next morning their camp was moved to the lake and preparations made for the hunt. Near by was an old deserted camp of the Chippewa Indians. Scattered all about were evidences that the red men had been successful in their hunts. Many moose and bear bones were seen hanging to trees, where they had been placed by the Indians to propitiate some god of the chase. The shoulder-blades of the animals were striped with charcoal and adorned with strips of cloth and then hung to a limb as an offering. A number of fleshing-bones and other implements used by the Indians were found.

The purpose of this trip was to hunt moose, but it soon appeared to develop into a fight with mosquitoes for twenty-four hours each day. The experience of the first night in the swamp was repeated every night thereafter until the frost killed the insects. The ponies were sewed up in heavy blankets from head to foot, and smudge fires were kept constantly burning that the suffering animals might stand in the smoke. The tent was pitched and banked down so tightly on all sides that it was impossible for the insects to enter. Then with a lighted candle every mosquito inside the tent was tried by fire. It was stifling inside the tent, but it was torment on the outside. During the night the horses threshed around among the bushes, rolling, kicking, squcaling, and tumbling about in their efforts to rid themselves of the pests, and at last one of the animals got down and began rolling in the ashes of the fire. This proceeding was due to the fact that the smudge fire had burned low, and Dyche prepared to go out to start the fire again. Brown endeavoured to dissuade him, but it was necessary to relieve the horses or they might tear down the tent. Covering himself, head and hands, with cheese-cloth, the naturalist went to the remains of the fire. The air was thick with mosquitoes. The warm blankets of the horses were literally hidden by the insects. The tent was so completely covered that not a sign of white could be seen anywhere. The droning noise changed into a screaming sound, rising and falling with the movements of the swarms, which extended for miles upon miles in every direction. Dyche was at once covered. After starting a fire with the old logs, so that a dense smoke would rise, he returned to the tent, but there went with him hundreds of mosquitoes, and another hour was spent with the candle getting rid of them. This experience was repeated night after night, while the days were spent in brushing the swarms away from the nose, mouth, and eyes. With every precaution it was impossible to be free from the bites, and their hands and faces were swollen from the poisonous stings. The tops of the tall tamarack trees were the only places where the hunters could be free from the pests.

An old Norwegian lived in a small log hut near, and the hunters paid him a visit. They soon discovered that the old fellow had evidently lost his sense of smell. He spent his time in catching and curing fish, which he caught by means of a dam and fish-trap in the stream. These fish were hung on a pole to dry, and with them hung the skin of the old man's dog, which had died a few days before. The Norwegian offered some fish from the pole to his visitors, but they were declined with thanks. The extent of his conversational powers was limited to "Ja, so," which fact gave him the name of "Old Ja So."

The days were slipping by and neither moose nor signs of moose were found. Dyche had come to the country early in order to get moose calves, but as time passed he saw no prospect of accomplishing his object. Days were passing into weeks, and something must be done soon if a specimen of a calf was to be secured. Finally it was determined to move camp to the other end of the lake. They were just ready to proceed, when the horses took it into their heads to add to the troubles which had followed the party ever since they had come into the country. The load had been securely packed and then tied down in order to keep it in place on the wagon. Just as they were ready to start the horses made a bolt and were off towards Loughridge's ranch at full speed. Dyche and Brown ran after them for a few hundred yards and then gave up the chase and proceeded slowly on the broad trail left by the bounding wagon. The outfit was soon out of sight and the naturalists followed at a fast walk, wondering what would be left by the time the horses had finished their escapade.

The trail was almost lost, when a horse-blanket lying in the swamp showed them the way, and then they found a ham. With a blanket and a ham they could at least sleep and eat, so with this grain of consolation they continued the chase. Here the wagon had struck a root and there it had almost turned over. A piece of the tail-gate was found and then a broken bit of the wagon-box stood up in the swamp. Here lay the lid of the "chuck-box" and there the top of the cracker-box. Hurrying on, they finally came in sight of Loughridge's ranch, and there, standing at the corral fence, were the runaways, apparently asleep. Old Buck had been warranted as one of the gentlest horses in the country, but his gentleness was laziness, and it turned out that he would run away at every opportunity.

Dyche's work in the taxidermic shop stood him in good stead now, and with some bolts and strap iron found in the remains of an old mowing-machine he soon had the wagon in as good a condition as before. The hunters now proceeded to the head of the lake and renewed their hunt for a moose calf. They were in the midst of the great swamp and they explored the country for miles in every direction. This was surely the home of the moose and there were signs that the animals had been there, as old beds and tracks were seen. Dyche soon discovered that while it was easy for him to find his way about the mountains, it was a different matter here. He spent one bad afternoon searching for camp, under the impression that he would have to remain in the swamps all night with no protection from the mosquitoes. It was only by the greatest luck that he finally came to the river and followed the stream to camp. Day after day was spent in unsuccessful hunting, and the hunters were almost ready to give up in despair, when they discovered a place where a cow moose and her twin calves had made their beds. The beds were more than a week old, but they renewed the enthusiasm of the hunters.

Camp was again moved and they went as far as possible into the swamp. The whole country was covered with "moskegs," a peculiar formation composed of plants and grass which had become so matted as to make a new soil on the top of the water of the lakes, and finally had become so firm as to give support for the growth of trees and shrubs. There were hundreds of acres of cranberries and blueberries in the swamp, and while to all appearances the surface was firm ground, it was in reality a most treacherous place, for at any moment the hunter might sink to his hips in water. Here were found so many evidences of moose that the hunters were satisfied that they had at last reached the right place. All was going well and the indications pointed to a successful moose-hunt, when Brown had the misfortune to let his axe slip and cut his foot between the heel and ankle. Dyche had gone for wood, and when he returned he saw his companion rolling on the ground and moaning. The flow of blood could not be stopped for several minutes. The wound was dressed, but Brown was compelled to lie still for several days, while to Dyche fell the greater part of camp work and all the hunting.

East from the camp there stretched a great meadow, and it became a daily habit for one of the hunters to mount the wagon and examine this open place for any possible indication of game. After Brown's foot had begun to heal and he was doing light work about the camp, he was dissuaded with difficulty from going out to hunt. One morning Dyche was standing on the wagon scrutinising the meadow, when he called to Brown in a low but excited voice. Brown climbed up as quickly as his lame foot would permit.

"What is it?"

"There, don't you see them? Just at the edge of the timber."

Brown eagerly scanned the country with field-glasses and at last caught sight of the animals.

"Moose!" he exclaimed; "they are moose, sure. They are nearly out of sight behind that point of brush, but they are moose, sure."

Brown crawled down from the wagon, and despite all warnings from Dyche began to make preparations for a hunt. He caught up his Marlin and Dyche his Winchester, and off they went, Brown hobbling along on his lame foot while Dyche tried in vain to induce him to go back to camp.

"Get back. You'll ruin that foot," said **Dyche**.

"Guess I can stand it," was the only answer.

"You'll catch cold in it in the water."

"Guess I can stand it."

The animals had moved south into a bunch of willows which extended into the swamp. The wind was in the northwest and the hunters made a circuit to the south, hoping to head off the moose before they got out of the country. Brown, whose foot seemed to give him no trouble, was of the opinion that they would not be able to get ahead of the animals unless they stopped to feed in the bunch of willows, for when they once started they usually kept going for many miles.

Reaching a point three-quarters of a mile to the south of the place where the moose had been seen, the hunters thought it time to look for signs. There were no indications that the game had passed that way. Dyche climbed a tall tree and scanned the country in every direction, but there was no living thing to be seen. A long consultation resulted only in the belief that they knew that there were three moose in the country, and the hunters made their way slowly back towards camp. Dyche thought the animals might be feeding in the willow thicket, but Brown said it was hardly possible. His experience taught him that they always fed away from the lakes and rivers. Their beds and other signs were nearly always found far in the woods, which would tend to carry out the theory. He felt quite sure that the moose were miles away to the south by this time. While giving due weight to Brown's theories, Dyche could not get rid of the idea that the animals might be feeding in the willows, and at last Brown said he would crawl through while Dyche went around on the edge to shoot if the animals rushed out. Brown was cautioned not to shoot the old cow, for the season was so far advanced that her hair would be in bad condition and her meat was not needed.

"Don't kill the cow, but if you get the two calves I'll give you ten dollars extra," was the parting injunction.

"They are dead calves if I see them," was Brown's reply, as he crawled into the brush.

Five minutes of silence; then a shot rang out, quickly followed by four others in rapid succession. A bullet sang close to Dyche's head and he dropped to the ground to escape the next one. A whining, bawling noise was heard and then Brown shouted. Hastening to the spot, the naturalist found Brown bending over a moose calf, not yet dead. A short distance away lay another.

"Brown, you've done well, but I ought to have been in it and got one of them."

"That's so. These are your moose."

"Where's the old cow? Which way did she go?"

"Go? Why, she was lying down there in the edge of the brush and got up within five rods of me."

"Well, where did she go?"

"Go? Why, she just stood there and I shot her for her impudence. She looked as if she was going to charge me. I never yet allowed a moose to get up within five rods of me and run off if I could help it."

The hair of the cow was found to be in excellent condition. It was short and thick, almost black. Her skin and skeleton and meat were saved after careful measurements, and they were taken to camp with the calves next day, while traps were set about the remains of the dead cow. The calves were an iron-gray colour, shading into black on the under sides and lighter on the legs.

A week was spent in getting the specimens in order, another in returning to Warren, and a third in packing the specimens and preparing them for shipment. Old Buck was traded off for an active little black horse and then a trip was taken to Thief River Falls, where a number of moose calves were in captivity. A day was spent in examining them and making notes and drawings. Some of the calves were so tame that they allowed themselves to be measured with a tape-line.

CHAPTER XVI.

In the Swamps—Habits of the Moose—The Moose-Call—On the River—Good Shooting Secures a Group—The King of Game Animals—The Naturalist nearly Killed.

SEPTEMBER was half gone. The cold north wind brought ice and snow and the conditions were changed in the swamps. Travelling was as difficult as it could well be. The chill air and snow put a damper on the mosquitoes, however, and while these pests were not exterminated they made their appearance only on the warmest days. Water was everywhere and the whole swamp seemed converted into one vast lake. It became necessary to build a platform of poles and logs on which to make camp. On this platform, twelve feet wide by twenty-four long, was the equipment of the hunters. One end supported the tent, while on the other a little clay and wet débris from the bottom of the swamp was placed. On this clay bed the fire was built, and as it burned down to the water's edge the clay and ashes made a good foundation. To prevent accident, however, the fire was extinguished whenever the hunters left camp.

All the hunting was now done in water, and the

continual wading placed Dyche's knees in a condition which strongly reminded him of goat-hunting in the Cascades. **Higher** and higher rose the water, and the horses were in danger of losing their hoofs from standing in so much moisture. A quaking-asp grove on a sand ridge was discovered about a mile from the platform camp, and it was decided to move to this. Three days later the hunters were ensconced in a new camp. Water or no water, they **were** determined **to stay** until they secured **a moose, even if it** took them until Christmas to **do it.** With a ton of good hay cut for the horses, there was nothing to prevent an enjoyable hunt in the water.

Cold nights with heavy frosts had changed the aspect of the whole country. The shivering aspens had dropped their leaves, while the tamaracks were changing to a golden brown and covering the earth with their needles. Vast armies of ducks, geese, and other aquatic fowls covered the swamp in every direction and streamed overhead in countless thousands all day and night, with their constant "honk, honk," like the bugle-calls of some great army. Many flocks of willow-grouse came about the camp, and their cackling could be heard in all directions in the cold frosty mornings as they sat on the bare trees or the upturned roots of some overthrown giant of the forest. Thousands of incidents new and strange to the naturalist were of daily occurrence and lent zest to the hunting.

But the horns of the bull moose were now getting hard, and **it** was a big bull moose that Brown and Dyche were after. The naturalist wanted a very large

one to head his group. Brown was anxious to please
him, but he shook his head whenever this big fellow
was mentioned.

"You may get one, but I doubt it. I've been hunting moose for five years and have never yet seen a
very large one. Big bulls are mighty scarce. I
found the horns of one once having a spread of forty-
eight inches, but I think he was the last of the
lot."

While Brown was inclined to be discouraging,
Dyche was determined to have a big moose if
there was one left in the swamp. The season for
calling moose had arrived, and while the naturalist
had heard much of this manner of attracting the animals, he had never seen it done. Brown was an old
moose-caller and had brought the animals up to him
and shot them. He had learned the art of an old
Nova Scotia moose-hunter who had spent a lifetime
at the business. Dyche took lessons until Brown
pronounced him proficient enough to deceive the oldest moose in the woods, and then they were ready to
go after the big fellows.

"The moose knows what he's about every time,"
said Brown. "You can't fool him unless you do
everything just right. A moose can smell where
you have passed along several hours before, and
whenever he gets scent of a man it's good-bye, Mr.
Moose. You can call the moose and get the answer
all right enough. The old fellow may be three or
four miles away, but he will come promptly to the spot
whence the call comes. He won't come in a bee-line,
but he will get there. He goes in a big circle around

and around the place or zigzags back and forth, gradually getting closer and closer until he is within a few hundred yards, and then he will go slowly and be sure that everything is just right. If there is the slightest wind he will scent you, and off he goes to some secluded spot ten or fifteen miles distant. The best way is to get on the bank of a lake or river, with the wind, if there is any, blowing towards the water, and then the moose will go backwards and forwards, quartering from one side to the other, until he is close up. If you make a sound, take a step, or move your cramped legs, all your trouble is for nothing, for the old fellow will never stop to investigate, but will go out of the country as fast as his legs can carry him.

"You are in no danger whatever from the moose, for they never attack a man unless badly wounded and unable to get away. But there are many strange things that happen while you are lying out in the swamp waiting for the moose to come. Sometimes a big fellow will be answering your call and coming towards you, and while your attention is directed towards him you will be surprised by the appearance of a young bull which has also been attracted but which has come silently for fear of the big one. The young fellow is not saying much, but the first thing you know he stumbles in on you and spoils the whole thing.

"Sometimes the bull that answers will have a cow with him. When he hears the call he will leave his cow and start for the new one. The old cow gets jealous and starts too, and while the old bull is threshing around in the bushes she comes straight to you ready to fight her rival. Then again I've known bears to come where they hear the cow bawling. I suppose they think there is a calf somewhere in the bushes and expect to get some of the meat. You may be lying in the bushes listening to the guttural grunts of the big bull moose as he makes his circuit around you, when suddenly a bear will appear in front of you looking for something to eat. Well, you ain't hunting bear just then and the sight is anything but pleasant.

"When you are calling moose and are answered by the barking of wolves or coyotes, you can just put up your horn and go back to camp. There will be no moose that night. I don't know why, but that is what old man Thomas told me and I have always found it true. The howling of wolves in reply to a moose-horn seems to be a sure sign that the moose will not answer."

This talk had the effect of making Dyche think that moose-calling was anything but a pleasant operation, and when he came to try it he found that the reality was even worse than the anticipation. One night when the wind had died down the first attempt to beguile a moose was made. It was wet and cold, and after Dyche had been stationed in a bit of swamp at the edge of the river, where he immediately got wet to the skin, Brown made a call. The first call was low, with the horn turned towards the ground, while the mouth of the instrument was muffled with the hand. After waiting fifteen minutes without an answer the call was made again, this time louder with the mouth of the horn free. Twenty minutes passed and no answer came. A third time the call was made, loud and clear, but no answer came. Again the call was given, this time with the full force of the hunter's lungs. It rang out over the forest like a trumpet.

With the birch-bark cone at his lips, the mouth towards the ground, Brown emitted a peculiar grunting squeal, and as he called he elevated the horn with a spiral motion high in the air over his head and then back down towards the ground, the sound rising and falling and vibrating through the forest. This was repeated three times and then they waited in the stillness for the answer. The noise is the bawl of the cow moose, and is a combination of the prolonged howl of a dog and the lowing of a cow, if this can be imagined.

The sound penetrated the depths of the forest for miles, and the echoes had barely died away when there came from a far-distant part of the swamp a most peculiar grunting or thumping noise. It sounded something like the rapidly repeated "woof, woof" of a big dog. The sound was continued, the moose gradually drawing nearer and nearer, all the while uttering his deep guttural grunt or bark. It was his answer to the cow, and while he was undoubtedly three miles away when he first heard the call, he went directly to the spot where the two men were lying concealed, almost frozen from their long wait in the frosty swamp grass. Dyche was stiff and numb from his hips down, while Brown was in little better condition. The old fellow was coming, however, and they did not mind the disagreeable part of it if they could only get a shot at a large bull moose. The "woof, woof" came closer, and at last they could hear the animal threshing around in the bushes about a hundred yards away. At last he appeared at the edge of a thicket about a hundred yards away from the hunters. He was threshing the bushes in an extraordinary fashion and all the while keeping up his grunting noise. As he circled around he could be seen as he passed from one clump of bushes to another. Finally he came out into the opening and trotted about forty yards forwards towards Dyche and Brown. Here he stopped stone-still, with his head erect and his great black body looming up like that of an elephant. Dyche whispered that it was time to shoot, but Brown objected, saying that the moose would come closer. But his prediction was erroneous, for the bull suddenly wheeled about and dashed into the bushes, where he resumed his grunting and threshing. For over an hour he continued tramping around within a hundred and fifty yards of the hunters, until at last Dyche grew so cold that he could not have hit a moose ten yards away. No noise had been heard for ten or fifteen minutes, and the naturalist said he was going back to camp. Brown gave another muffled call, and as no answer came they started for the tent. Just as they were well under way the moose gave another grunt and began beating the bushes with his horns again. The hunters crouched down and waited for another half-hour, when Dyche made a break for the camp, declaring that there were not enough moose in the woods to keep him there another minute. They reached the tent about midnight, chilled through.

Several inches of fresh snow fell soon after, and Dyche made a circuit through the country. He found fresh moose-tracks and started after the animal, following the trail for about eight miles without seeing the moose. Great holes had been pawed in the snow and the trees and bushes showed marks where they had been threshed with the horns of the bull. In several places the moose had stopped **to feed upon the tender** tops of willow bushes and the red osier or " **killikinic.**" Sometimes the trail led straight through the centre of these patches of willow and osier, but usually it skirted them. The moose seemed to enjoy going through the centre of the spruce and tamarack groves, where there was plenty of water, which gave Dyche a pair of **very** wet feet.

One evening Dyche concluded to try calling alone. Taking the birch-bark horn he went to the bank of the river, about four miles from camp. **The wind went down with the sun, and** just as the great yellow disk disappeared the **naturalist gave a muffled call.** No answer came and the call was repeated. Far off in the tamarack swamp a sound was heard which the hunter at first failed to recognise, although he finally concluded that it was a moose. His supposition was not amiss, for soon the animal was heard coming directly towards his place of concealment. The moose was on the opposite side of the river and at last he could be heard in the bushes, threshing around with his immense horns until one could almost imagine that a dozen bulls were fighting in the **forest.** Dead limbs would be knocked down and his horns would rattle against the trees, and then all would be silent for several minutes.

The moose was between fifty and seventy-five yards away, and now Dyche began to fear that the animal would refuse to come out where he could be seen. All the while the naturalist was getting colder and colder, but he did not dare take a step or get into an easier position for fear of sending the animal back to the forest. It became evident that the moose did not intend to cross the river, and the hunter determined either to bring the game out or send him back home. With the birch-bark horn he gently rubbed the top of the willows. It was the challenge from one bull to another. Instantly all sound from the bull ceased. The big moose stood and listened. Dyche waited fully half an hour, but never a sound came from the bull. The rain now came down heavily, and Dyche decided that it was of no use to waste more time with the moose. He gave the bushes another rub and then emitted a defiant grunt from the horn. The effect was magical. The moose bolted straight for the deep woods, making as much noise as a pair of horses running away with an empty hay-wagon. He knew he had been deceived and stopped for nothing. Dyche was half-frozen and his legs were stiff, but he was able to get a little amusement out of the bull's frantic flight.

For three weeks the swamps were hunted in every direction, and during the whole time not a gun was fired. An old flat-bottomed scow which they had brought in with them was caulked, and they started on an exploring expedition up Moose River. Moose were undoubtedly in the country somewhere, for they answered the calls whenever they were made. They poled along up the river, intending to stop morning and evening **and** call for moose **when** the weather was suitable.

It was hard work, but they pushed on, cutting away the overhanging **boughs** and shoving aside the driftwood with their poles. About the middle of the afternoon of the first day they reached a place where the river widened into a **small lake.** Just as they rounded the point into the little lake a moose calf was seen standing in the water. The calf started through the brush to the woods and Dyche shot as it was disappearing in the thicket. Again and again he fired, sometimes at the calf and sometimes at the woods, but with no evidence that the wild flight of the animal was impeded. Just as the last cartridge in the gun was fired another calf, which had been standing in the water and brush farther up stream, started out, and the naturalist seized Brown's gun and sent a shot after it. The boat was pushed ashore and search **was** made for the calves, but the only wounds apparent were those made in the trees. There were signs that three moose had been feeding near by, but the animals had evidently gone elsewhere on urgent business.

They now pushed their **way up the** river and at sundown Brown gave several calls and got an answer. The grunts continued several hours, but as they did not appear to come closer the hunters crawled into their sleeping-bags, leaving the moose to grunt at his pleasure. Shortly after midnight Dyche was aroused by a strange noise, but the barking of a fox and a spirited conversation between two owls was all that broke the stillness, and he dropped back to sleep again. An hour later he was wakened by Brown, who was crawling out of his blankets.

"Get ready," came in an excited whisper from Brown.

"What for?" was the sleepy response.

"Don't you hear that noise? A big moose is coming."

The word "moose" fully aroused Dyche and the hunters prepared for action. The bull came on, making a great noise with his grunting and threshing, but after he came within fifty yards of the hunters he refused to come farther and remained just within the thicket and out of sight, all the while keeping up his noise. It was very aggravating for the hunters to sit all night in the cold, waiting for the moose to come into sight, and then, just when daylight appeared, to have him retreat to the place whence he came. They got into the boat, much dejected, and returned to the home camp, where a good night's rest placed them in condition for another trial. They spent the day in going up and camped at the head of Moose Calf Lake, as the place had been named by Brown in honour of Dyche's adventure with the calves.

They pushed on up the stream all next day and went much farther than they had ever been before. When evening came they allowed the boat to drift with the sluggish current and floated in this lazy manner until about ten o'clock, when they crawled out on the bank and sought a dry spot on which to sleep. As the evening wore on the cloudiness turned to a drizzling rain and then the water poured down. It was a dreadful night, and morning found two bedraggled hunters who were inclined to believe that the world is but a fleeting show. Tired and sore from their exertions in pushing the boat all day and weary and care-worn from a sleepless night in the rain, they allowed the boat to take its own course.

Occasionally one or the other would dip his pole in the water in a half-hearted way and then lapse into inactivity again. Thus they moved along until the afternoon, feeling that there was nothing to do but get back to camp and secure some sleep. A few big owls sitting on the tops of the trees eyed them in a quizzical manner and appeared to make sport of their dreary condition. Dyche was sore, mentally and physically, and he felt that moose-hunting was a failure, especially in a country where there was water not only under-foot, but overhead, all the time. The season for calling moose was about over, and they had succeeded in getting them to come no closer than the bushes where they could remain hidden. While these doleful thoughts were chasing each other through Dyche's brain, Brown capped the climax by saying:

"Yes, a big bull moose is the rub."

Slowly drifting along, they reached the spot where Dyche had had his experience with the calves a few days before. As they pushed on through the lake, Dyche glanced over into the bushes at which he had wasted so much ammunition, and Brown quietly remarked:

"Yes, that is the spot."

No comment was necessary on Dyche's part, as he felt that it was not his time to talk.

Just as they were making the turn at the lower end of the lake, Dyche, who was sitting in the bow of the boat, noticed a moose calf in the water about seventy-five yards below. Only the top of the animal's back showed above the water. As he reached for his gun he noticed a smile flit across Brown's face, and the calf turned towards him with an air that seemed to say: "Hello! is that you again?" In less than a second Dyche had his gun trained on the back of the calf and sent a bullet after it. The calf made a desperate lunge for the bank and disappeared in the bushes on the south side of the stream.

The boat was so unwieldy that the naturalist could not push it to shore fast enough, and he jumped into the water and ran for the bank, trying to get ahead of the calf. He ran as fast and as far as he could to get out into the open swamp, away from the brush which lined the shore, hoping to obtain another shot before the calf reached the spruce thickets. After running forty or fifty yards through the swamp, a noise on the north side of the river attracted his attention and he saw a big moose just going into the bushes. Dyche sent a bullet after it and then ran to get around a clump of bushes to a spot where he could catch another glimpse of the animal. A run of seventy yards brought the moose into sight again, and two more bullets were sent after her as she disappeared in the bushes. After another **run** around some brush he stopped to regain his breath, when a sight met his **eyes** which almost took away his breath.

A giant **bull** moose was just coming out **of the** bushes on **the** opposite side of the river, facing almost directly towards Dyche. In an instant it flashed through his mind that this was the animal which he had been looking for. It **was** evident, even at that distance, that the moose was a monster. His massive horns branched like the tops of the trees around him and his form showed that he was one of the ancients of the swamps. The naturalist realised that this was **the** opportunity **of a** lifetime. He had been hunting for this moose for three months and now he was here.

Bringing his Winchester to his shoulder, Dyche held his breath and commanded his throbbing arteries to be still. The bull was almost facing him and was just swinging around the clump of bushes, when, with the crack of the rifle, he gave a great lunge forward and then whirled in his tracks and started **back** into the woods. This **gave the** hunter time to pump in another cartridge, and a second shot was taken **as** the animal went into the bushes. Again a heavy lunge showed that the bullet **had** gone home, and as the animal went threshing through the bushes two more bullets were sent at the head, which could be seen going up and down as the bull struggled to get through the jungle of brush and logs.

The bull was now out of sight, and Dyche started to run to the bank of the river, where he could get a clear view across, **when the** calf jumped up from be-**hind a** log where it had been lying. The frightened animal ran fifty yards before Dyche could get a shot **at it, and then** it, **too,** disappeared and not a moose **was to be seen or a** sound heard.

Dyche ran back to the spot where he had left Brown in the boat, but **Brown had** gone ashore on the north side and the **boat** was drifting down the river. The naturalist waded to the scow and poled his way across. **He had** gone but a short distance when he heard a shout from Brown. Hastening to his companion, he found him standing over a mighty bull moose. He was, indeed, a monster. **His** great horns spread out above his head like immense shields, while his body showed that he was the giant of the swamps. Dyche was almost exhausted with his exertions and excitement, but he found sufficient breath **to** jump on **the** big animal and make the woods ring with the university yell. Brown looked at him in surprise and then started off to look for more moose. **Soon** his call was heard again, and this time Dyche found him standing beside the body of a cow. Dyche gave more cheers while his companion went to look for the calf.

Soon Dyche heard the vigorous use of strong language in the direction of the river, and hastening there found that Brown had undertaken a task which was almost too much for him. **The** calf, badly wounded, was standing in the water pulling back from Brown, who was holding **to one** ear while he braced himself by grasping the willows lining the stream with one hand. The calf was getting somewhat the better of this unique tug of war, and when Dyche reached the spot Brown was up to his waist in water and going deeper. Getting hold of the other ear, Dyche assisted Brown in landing the prize, which soon died of its wounds.

These incidents occurred in a very short space of time, yet to Dyche it seemed that he had accomplished the object of a lifetime. What a bull this was! A leader in every sense of the word and just the one to head the group which Dyche had in contemplation. The horns and the grand proportions of the body exceeded even the wildest dreams of the naturalist, and he almost feared that he would wake and find the whole episode a dream. He had seen many sets of horns, but never before horns like these.

The measurements were taken and the specimens dressed so that they would not be disturbed until they could be taken to the home camp. From the top of the back to the point of the hoof the big moose measured, just as he lay, eighty-five and a half inches; from the top of the back to the bottom of the hoof, eighty-one and a half inches. The standing height of the animal, after all corrections were made, was just seventy-eight and a half inches. This was equal to a horse nineteen and a half hands high, and above this towered the massive head with its wide-branching horns. The skin weighed one hundred and thirty-five pounds after it had dried one day in the open air. The heart was twenty-two inches in circumference at the base and ten and a half inches in length, weighing, when free from all blood and arteries, six and a half pounds.

The old cow and calf were fit companions for this monarch of the woods, and the three composed as magnificent a group of the finest game animals in the world as could be found in years of hunting.

Much has been said and written about the pleasure of hunting the buffalo, and that animal has been looked upon by the average lover of sport as the embodiment of perfection in the way of game. **He has** always been considered the king of the plains, **and** to hunt him has been the ambition of kings. His decadence has been mourned as the passing of an era in the history of the sporting world. While it is true that the bison, when he roamed the prairies in the countless numbers recorded by Lewis and Clarke in their " Explorations," afforded much excitement to the lovers of the chase, yet the fact remains that he had **none of the** characteristics which belong to the true game animal. The stupid beasts would stand in droves while the pot-hunter crawled up and shot down hundreds from a place of concealment. He was **an** easy victim of the red men with their simple weapons of the pre-Caucasian times and was the great food-supply for the natives. The body of the bison is large and affords food for many, while his robe gives warm covering, but he was never, in any sense of the word, a game animal.

But how is it with the moose, the giant of the swamps? Here, indeed, is the greatest of all the game animals of the North American continent. He is a monarch of the forests, and in addition to his size he has sufficient speed and cunning to outwit the wariest hunter, while his courage is equal to a defence to the death when he is too closely pressed or receives wounds which stop him in his flight. His home is in the swamps, where his cunning and instinct teach him that his enemies are at a disadvantage. His eyes are of the keenest and note any casual change in the appearance of the landscape, while his nose, composed of a combination of cartilaginous boxes, detects the faintest odour left by a passing enemy even hours after the trail has been made. He does not stand and investigate strange sights or examine unusual scents, but at the first indication of danger he flies to the deepest woods of the swamps and leaves many miles between him and the place where there is a possibility of danger.

The hunters had now been away from the home camp for three days and they were uneasy about the welfare of the horses, which had been picketed out where they could get grass and water. As they approached the camp they caught a glimpse of a man going from the river to the tent. Fearing that bad news from home must have sent this messenger into the swamps, the naturalist hurried forward to greet the visitors. Superintendent Wood, of the Pembina Farm, near Warren, and "Holy Smoke" Bolton had come to pay the hunters a visit, and finding the camp deserted they had taken possession to await the return of the owners. They had had a tiresome trip and tedious search for the camp, and had arrived just after the hunters had gone on their last trip up Moose River. They would have passed the camp had it not been for the little black horse which Wood had sold to Dyche. The superintendent had seen the animal feeding in the brush, and supposing it to be a moose had made an hour's stalk and discovered his mistake just as he was about to fire at the animal. The picketed horse showed the visitors that they were near camp, and they soon found the place and made themselves at home.

The goodly supply of home-made bread, jam, cookies, pies, and the other unwonted delicacies brought by the visitors was greatly relished by the naturalist and his companion. Bolton could only stare, while his ejaculations of "Holy smoke!" at the rapid disappearance of the good things showed how he came by his nickname. The three days in the swamps had put an edge on the appetites of the hunters that caused an exceedingly rapid diminution of the stock of provisions. But while exertion had improved their appetites, it must be said that their personal appearance had sadly deteriorated, and their visitors were constrained to make many remarks on the subject of tramps.

As the visitors found little evidence that the hunting had amounted to anything they were disposed to make light of the efforts of the naturalist, and many were the humorous remarks which were passed at the expense of men who would stay two months in the swamps and have nothing to show for it. The hunters contented themselves with silence, but next morning when they took their visitors up the river to the place where the three moose were lying it was their time for laughing, and Wood and Bolton confessed that the trophies were well worth the time spent in capturing them.

Three days were spent in getting the skins, bones, and meat to camp. Bolton and Wood were ready to return to Warren, and they took a wagon-load of meat out with them to the settlement, where it could be made use of. Bolton was enamoured of the camp life, and loading the wagon with fresh supplies returned to the camp in the swamp.

Hunting was continued with varying success until the cold November winds froze the swamps solid, and then the specimens were packed and the party started on their return to civilisation. All went well until the party was nearing Loughridge's ranch. Bolton had been driving all day and called to Dyche to relieve him while he walked a little to get warm. Just as the naturalist took his seat the horses started and the off wheel of the wagon dropped into a deep rut, while the other side struck a rock. The result was that Dyche lurched forwards, and before he could recover himself he had fallen to the double-trees and thence to the ground immediately in front of the wheel, which passed directly across his abdomen, the heavily loaded wagon almost crushing the life out of him. As the front wheel went over him he raised himself so that the hind wheel struck his shoulder and threw him forwards under the wagon, out of the way of more danger.

The accident was serious and the naturalist was in such pain that he could not bear the jar of the moving wagon. After lying at the ranch two days it was decided to leave the camp effects and specimens at the ranch and make an ambulance of the spring wagon in order that the nearest surgeon could be reached. The trip to Pembina Farm was made, and two weeks' rest there so improved the condition of the patient that he concluded to go back into the woods instead of following out his intention of returning home at once. A team was hired to bring in the specimens from Loughridge's ranch, and after these were packed and shipped home, Dyche made his arrangements for a trip to the Lake of the Woods with Brown and a hunt with the Chippewa Indians.

CHAPTER XVII.

With the Indians—How Indians Hunt Big Game—The Parallel Trails—Indian Superstitions—A Potent Beverage—Moose all "Nickoshin"—Return to Civilisation.

IT was a lonesome road which the hunters travelled to War Road River. Ice covered the swamps, but it was not firm enough to support the heavily laden wagons, many of which were passed sticking fast in the mud, their owners patiently waiting for a hard freeze, so that they might get through the swamps on the surface. Every camp was full of moose-hunters who had strange tales to tell of their adventures and prowess, but little to show. They knew all about moose-hunting, and found such ready listeners in Dyche and Brown that they were disposed to "talk large" for the benefit of the strangers. The naturalist and his companion pushed on and finally reached Brown's little cabin, near the village of the Chippewas. Dyche was still very sore from the effects of his fall, and while Brown busied himself in cleaning the cabin and making it habitable, the naturalist endeavoured, through the medium of an interpreter, to make arrangements with the red men for a hunt. The Indians flocked to the cabin to see the new-comers, but they were suspicious of all white

men, for their experience with traders had taught them that white men were not all honest.

Dyche used every argument that he could bring to bear, but with no apparent success. He even offered to pay full price for the game that he killed, as well as that killed by the Indians. He was willing to buy everything killed during the trip, and in order that there might be no doubt about it he offered to leave the money with any trader whom the red hunters knew. All his talk seemingly had no effect. The Indians were shy and made many signs to each other, grunting out their monosyllables and evidently not convinced that they could accommodate the naturalist. Finally the interpreter told Dyche that they were dry. "Too much talk, too little firewater." The savages puffed hard at their stone pipes when told that the hunters did not have any and did not use any.

As soon as Dyche and Brown finished their supper they began preparations for sleeping, as they were both tired from the day's exertions. The Indians took the hint and silently filed out of the cabin. By daylight the Chippewas were prowling around the cabin, and at the first sign that the hunters were awake, the red men crowded in and sat around the fire watching the preparations for breakfast.

Dyche spent the entire day in an attempt to reach an agreement with the Indians, but made little progress. The Chippewas were perfectly willing to hunt for him and sell him their game, but did not like the idea of his being with them on the trail. They could not understand why he desired to go along. The naturalist patiently explained over and over again that it was necessary for him to skin the animals and prepare the specimens. He told them what he desired to do with the game, and as Brown had mounted several birds here during one of his previous visits, the explanations were easily understood, but the red men were unwilling to give in. Their obstinacy was finally overcome, however, and they agreed that Dyche should go with them. Maypuck, the **chief**, Kakagens (Little Raven), Machiveness, and Gib, the interpreter, were to make up the party. The chief could speak a few words in English, but the interpreter was necessary.

While these arrangements were being made Dyche spent much of his time in visiting the Indian village and studying the home life of the Chippewas. The village was composed of log-cabins and tepees covered with coarse grass and birch bark. The principal occupation of the tribe was fishing, and the main food-supply was fish and a peculiar black rice found growing around the lake. The squaws made nets and prepared the fish and the skins of game while the bucks lay around doing much smoking **and** no work. The Indian is tireless on the chase, but he thinks his work done when the animal is slain **and he** leaves the labor at the camp to his squaw. Brown had been here often before, and he said that in the spring the Indians varied their diet of fish and rice with the eggs of aquatic birds, which bred in great numbers in the vicinity. The stage of incubation of the eggs cut no figure at all in the case, the Indians rather giving preference to those in which the young bird was about to break through the shell.

The village was alive with wolfish dogs, whose principal occupation seemed to be fighting against every other dog in the neighbourhood over the scant scraps left by the Indians. As the food refused by an Indian is barely good enough for a dog, the animals had a hard time in getting a sufficient supply to keep life in their bodies through the winter. While there was no danger of their attacking the hunters, yet it was necessary carefully to guard the camp to prevent their depredations.

Many of the squaws sat hour after hour on the ice of the lake, fishing, either with hooks and lines or with nets which they ran under the ice and left for two or three days at a time. When a pickerel or white fish was caught the woman jerked it out on the ice and killed it with a blow of an axe. These fish were cured and those that were not sold to the traders were eaten. Dyche walked to the edge of the lake to witness the fishing operations of the squaws, but as soon as he approached, the women gathered up their lines and blankets and made a dash for the village. Gib, the interpreter, laughed at the incident and explained their actions by saying that they were all young squaws and were afraid of a man.

Dyche supposed that all arrangements were now complete, but he was greatly surprised to find that the Indians made no preparations for departure. They now said that they thought he would not keep his agreement about paying for the animals killed by himself. There was another long pow-wow, and at last he was able to convince them that he would do everything that he had promised. He offered to leave the money with Brown or any other trader whom they knew, but just as the party started they decided not to trust the traders, but told the naturalist to keep the money himself and pay it to them when the animals were killed. With this understanding the party filed out of the village, Maypuck drawing the sled on which was packed the camp equipment.

Dogs are the usual draught animals of the Indians, and when Dyche asked why they were not used on this trip the response was:

"Dog no nickoshin [no good]. Too much brush."

The Chippewas found a malicious pleasure in travelling fast, apparently to try the mettle of the naturalist, whose side was still painful and who kept up with his red companions with much difficulty during the first few days of the trip. It was now very cold, the temperature going as low as ten and fifteen degrees below zero every night. During the day Dyche's mustache and eyebrows froze solid as he walked along. The Indians were warmly dressed in heavy clothing made from blankets, and Maypuck, in addition, had a peculiar hood of white flannel with a blue fringe running from the front over the top of the head to the neck. This hood was shaped something like the old-fashioned sun-bonnet and was very warm, and highly prized by the chief. As an especial mark of favour he offered to let the naturalist wear it, but Dyche was a little shy of its inhabitants and declined as gracefully as possible.

The Indians kept up a species of dog-trot all day over the moskegs, keeping out of the dense forests as much as possible. This trot was one means of keeping warm, and while it is trying to the muscles of a white man at first, he soon becomes accustomed to it and finds it an easy way to travel. Just about noon the Indian who was leading suddenly stopped and began preparations for building a fire. One of the others went a few yards away and began chopping a hole in the ice at the edge of the swamp. Dyche supposed that he was after water and took a small bucket to the place, but was surprised to see the Indian pulling a large piece of moose-meat from the hole. It had been placed in the water in the course of some previous hunt. The flesh was white on the outside, but inside it was good and sweet. The Chippewas said that it would keep a long time in the ice-cold water without spoiling. As meat is never spoiled for an Indian, this assertion can be taken in an Indian sense. Dyche discovered a bear's skull tied to a tree and started to take it down to examine it, but was prevented by the Indians. On this, as on numerous occasions when he asked concerning peculiar customs and habits of the savages, the only response given him was: "Oh, that's Indian." This was accompanied by a significant shrug of the shoulders and such an evident desire to avoid the subject that the naturalist did not press his questions.

About the middle of the afternoon Maypuck called a halt and the Chippewas made camp. A quantity of dry tamarack logs were cut and brought in to keep the fire going during the night. Nights are bitterly cold in this country, and the Indians evidently did not believe in the doctrine advanced by their brethren of more southern latitudes, "White man make big fire, go 'way off and freeze; Indian make little fire, stay close by and keep warm," for they made a roaring fire and kept it supplied with wood all night.

The Indians did little but eat and smoke while in camp. Even in the night they toasted pieces of meat before the fire and ate them. It was so cold next morning that everything in the woods was popping and cracking. Maypuck took Dyche and, with the thermometer marking twenty-eight degrees below zero, started after moose. The chief moved like a cat through the jungle of brush and fallen timber, but walked rapidly when he reached open places. After travelling in this way about three miles the Chippewa suddenly stopped and threw up his hand as a warning to his companion. Dyche stood perfectly still while the Indian slipped cautiously forwards. He worked forwards carefully and at last beckoned Dyche to come on. The trail of a band of moose was seen in the snow, and after examining the tracks closely for several minutes and looking at the tops of the willows which had been nipped off, the chief decided that the animals had passed along but a short time previously and had gone north.

Dyche now learned something in the way of hunting which surprised him, but which he utilised many times later on in his hunting expeditions. Instead of following in the tracks of the moose, the chief walked fully two hundred yards off to one side and then followed along on a line parallel with the trail of the animals. He moved very cautiously but very rapidly for about a mile and then crawled towards a bunch of willows, through which the trail evidently ran. So carefully did the Indian go that Dyche was sure there were moose in the willows, but this was not the case. The Indian was taking no chances, and he used every precaution just as if he was sure he was crawling up on the game. The bunch of willows and killikinic showed **many signs** of the recent presence of the moose.

Maypuck took a small stick, placed it in the trail and broke it, pointing to Dyche's foot and shaking his head. The naturalist understood readily enough that he must not break sticks on the trail. The chief now took up Dyche's foot and placed it directly in the track of his own moccasin, indicating that only one trail must be made, in order to lessen the danger of breaking twigs. Having thus cautioned **his** companion, the Indian renewed his stalking with **even** greater care than he had exercised before. As he approached another clump of willows Dyche was again deluded by the extreme caution displayed by the Chippewa, who crawled to the thicket as if creeping up on an animal in sight. When the Indian began examining the place Dyche went up to him and found three beds where moose had been lying but **a** short time before. As the moose had walked away they were evidently not frightened by the hunters.

Maypuck examined everything very carefully. His next move was a puzzle to Dyche. The Indian took the naturalist by the coat and pointed in the direction the moose had gone, and then at the tracks. Dyche could not understand. Maypuck sat on a log and smoked. After finishing this he again went through the coat-pulling operation, but his meaning was still hidden, so Dyche took the red man's coat and repeated the sign. Maypuck smiled at this, took another smoke, and said:

"Camp nickoshin."

Dyche nodded his head, after which the Indian smoked awhile and said:

"Moose nickoshin."

Again Dyche nodded his head in approval and the Indian smoked for a moment or so longer. Finally the chief put up his pipe, got up from the log, saying "Camp nickoshin," and started off in that direction. After a tramp of eight miles in an almost straight line, they reached camp just at dark. The naturalist now sought an explanation from Gib of the strange actions of the chief, and found that Maypuck was trying to find out whether he wished to follow the moose or return to camp.

The other Indians reached camp before dark, Kakagens being the last to come in. There was blood on his hands and coat and he reported that he had killed a cow moose. Next morning camp was moved to the place where the cow was lying, and the day was spent by Dyche and Brown in skinning the animal and caring for the specimen. The moose had frozen solid, and it was necessary to carry the legs to the fire and thaw them out before they could be skinned.

Dyche found that a small hole had been cut in the neck of the cow and a piece of tobacco had been inserted therein. Gib said this was always done, but when the naturalist asked for the reason and why a piece of the moose was decorated with rags or streamers and hung to a tree, he received the usual shrug of the shoulders and the reply that it was "Indian."

The party spent a week in this locality, but it was so bitter cold that little game was found. The thermometer marked twenty-eight degrees below zero nearly the whole time that they were in the woods. Many small fur-bearing animals, such as martens and foxes, were killed, and the Indians appeared fond of the flesh of the former, but Dyche found that it was very tough and tasted like an old rabbit. In addition to his other apparel, each Indian had a robe made from the skins of the white snow-shoe rabbits which were found in great numbers in the woods. The skins were cut into strips about an inch wide, the strip being cut continuously around the skin, making one string. These strings are then made into a triple pleat, or braid, and these braids are sewed together as thrifty housewives sew carpet-rags into rugs. This is done until the robe is as large as an ordinary blanket, and the Indians then have a warm robe in which they wrap themselves at night.

The evenings of Indian hunters are spent much after the manner of white hunters; stories are told and jokes passed, while laughter is frequent and prolonged. All the while the moose-meat is not neglected, and every man in the party helps himself to delicacies. From an Indian point of view the finest tidbit of the moose is a piece of intestine, nicely cooked by boiling. The intestines are stripped and placed in a pot, where they are boiled until they assume the appearance of huge wrinkled snakes. The chief then passes the dish around, apportioning a share to each member of the party. Dyche was honoured with the part which was considered the *bonne bouche*, and after he got it he was in a quandary. He did not like to refuse, for he feared he would insult the chief, yet he could not coax his stomach to receive the morsel. He had eaten many Indian dishes much against his will, but he felt that the time had come to draw the line. He thankfully accepted his portion and then began to devise a way of escape. Cutting a stick, he inserted it into the delicacy and placed it before the fire to roast, telling Gib that he preferred it that way. He was in no hurry about it, and by the time it was well done the Indians had finished eating and had scattered through the woods. Now came the longed-for opportunity, and as the last Indian turned his back the stick slipped and the obnoxious piece fell into the fire, where it was soon reduced to ashes.

As provisions, excepting moose-meat, were almost exhausted, it was decided to return to the spot where the toboggans had been left at the first camp. Moose-meat is good eating, but the white men found that it became very distasteful when there was nothing to go with it. Indians eat their meat without salt, but when they discovered that Dyche carried a little bag of salt in his pocket, the red men became persistent beggars for it. In fact, the naturalist discovered that there was nothing they did not want and beg for when they saw others have it.

When the packs were adjusted and the party was ready to start back Maypuck motioned to Dyche to lead the way. The latter shook his head, but the Indians stood back and waited. Gib explained that if Dyche did not lead they would not go back with him. Thinking that he had another superstition to contend with, the naturalist took the lead and started off at a rapid gait. The Indians carried heavy loads, and as Dyche took the usual dog-trot on the trail the red men followed closely in his tracks. Dyche was a little puzzled at first, but by careful watching he saw sufficient signs of the old trail to convince him that he was taking the right direction, and at last, when he saw traces of one of the camps at the edge of a moskeg, he began to increase his speed. As the Indians had had so much sport with him when they first left the village, he concluded to return the compliment and show what a white man could do when he tried. Faster and faster he went until he was almost running, with the Indians stringing out in single file behind him.

Looking back occasionally, he noticed that his red companions were strung out for fully half a mile, Maypuck being the nearest, two hundred yards away. The sight so encouraged the naturalist that he put forth an extra burst of speed and soon had the satisfaction of seeing a clear track behind him with not an Indian in sight. He dashed across the clearing at full speed and reached the camp with only Maypuck showing at the edge of the woods, three hundred yards away, and coming at a dead run. Dyche quickly started a fire and began heating water in his tin cup. When the chief reached him the water was hot, and the hunter poured into it a quantity of extract of ginger. This he drank and then handed the bottle and cup **to the** Indian, who came up panting and perspiring. The chief gave a grunt and threw his pack on the ground. Taking the proffered cup and bottle, he set the cup on the ground and turned the bottle to his lips, swallowing half its contents at one gulp. He returned the bottle with a sigh and said:

"Nickoshin."

Dyche explained by signs that it was "nickoshin" for Maypuck **but** not for the others. The chief understood and grinned, and said not a word about the bottle when the rest of the party came stringing in. As a reward Dyche afterwards allowed him to finish the bottle. Dyche no longer wondered at the ability of an Indian's stomach to digest cartilages and even small bones, for a stomach that could stand half a bottle of strong extract of ginger at one swallow could digest almost anything. The Indians all reached camp at last and fell to talking and gesticulating about the wonderfully fast walking of the white man. Not one of them showed the least sign of displeasure at the incident, but, on the contrary, **they** seemed to think more of Dyche for his great endurance and speed.

The party had now been out two weeks, and it was decided to return to the village. The toboggans were loaded and hauled within ten miles of the lake and then unceremoniously left standing on the trail, while the Indians went on to their homes and sent their squaws to bring in the spoils of the chase. Dyche **now** spent his time preparing his specimens for shipment, and meantime he was constantly surrounded by an admiring crowd of Indians **of** all ages and conditions. Dyche was not disposed to drive them **away,** for he obtained much valuable information regarding the habits of the moose. It was a new and queer experience for the naturalist to sit with a dozen Indians and Brown and talk of hunting adventures through the medium **of** an interpreter.

The moose is the largest member of the deer family and the most awkward in appearance, but at the same time he is the wariest and shyest of them all. An old cow moose, with her ugly head and long ears, looks so much like a mule that new hunters frequent**ly let it** go, thinking they have run across some trader's stray animal. When they discover their **mistake** they make up for it by shooting the next mule they see, under **the** supposition that it is a **moose.** One of the most difficult anatomical constructions that Dyche ever attempted to mount was the nose of a moose. It consists of such a complication of cartilaginous boxes and partitions as to present many hard problems in taxidermy. The general colour of a moose, when seen at a distance, is black, but on closer inspection it is found to be tinged with red, brown, and gray hairs, the black changing into gray and white on the legs. There are many colour varieties, some being very dark while others present an ashen appearance.

Brown told much about the moose that he had learned from actual experience, while Dyche gave information gleaned from books, supplemented by many of his own adventures. This talk was interpreted **to** the other Indians by Gib and the red men were gradually drawn into the conversation, and they told strange tales of the habits and characteristics of the big animal.

"I saw an article in a paper the other day," said Brown, "which gave pictures of the European elk. That animal looked just like our moose. Do they call moose, elk, over there?"

"Yes," was Dyche's reply. "The European elk corresponds to our moose. They are very near cousins, but our moose is larger and averages darker in colour. The horns of the American moose are said to be more palmated than those of the European or Norwegian elk."

"Moose are found all through the northern part of the country, are they not?" asked Brown.

"They formerly ranged over a vast tract of country, from ocean to ocean and from the northern border of the United States to the Alaskan line, where he is still common along the Yukon River and many of its tributaries. They feed on the leaves and small twigs of the trees. I never saw grass or moss in the stomach of moose."

"Moose eat coarse grass sometimes," interjected Maypuck, **who** had been following the conversation closely.

"Maybe that's so," said Dyche, "but I've never **found** any grass in the stomachs of any that we have killed. In the summer they eat leaves and tender branches and in winter they nip off the ends of limbs of willows and birch and sometimes strip off the bark. The contents of the stomach of a moose in winter looks like sawdust. They may eat sparingly of the coarser grass, but I have never seen any indications that they do."

"There's one thing sure," said Brown; "they are the wildest animal that a man ever attempted to hunt."

"White man don't know how to hunt," said Maypuck. "White man scare moose away; Indian crawl up and kill him."

"The great trouble appears to be to get within shooting range of the animal. I have followed them through the snow many times, and always found that they had doubled back on their track and lain down where they could see anyone coming on the trail. When I got back to where the animal lay hidden it saw or smelled me at once and went out of the country. Indians do not seem to have any trouble getting them, yet they are not good shots and are armed with old shot-guns usually, which a white man would not carry with him. I know that their weapons are not good for over forty yards, yet they get the game just the same."

"Indian know how to hunt him," said Kakagens. "Indian no follow trail close like white man. Indian go 'way off and crawl up close. See moose lay down, shoot him in side. Moose smell good, see good. He walk way wind comes, walk back on trail, lay down see white man coming. Indian no follow, he go round, crawl up at side and shoot moose. White man in a big hurry and scare moose. Indian hunt him two, three days."

"I guess that's so," said Dyche. "I found that their beds were always made where they could look back on the trail, through an open place, with the wind at their back. They appeared to have walked back on the trail to where they could see a man if he was following them and smell him if he came up the other way. This nose, which has been puzzling me so much, appears to be made for the express purpose of smelling everything in the country. Sometimes, however, the moose gets ' rattled ' and **does** not seem to know which way to run. Sometimes they will stand until they are shot at three or four times. How many calves have you seen with one cow, Brown?" asked Dyche.

"I have seen many old cows with two calves and the Indians have told me that they have seen cows with three calves. The young cows have one calf, but the older and larger cows generally have two. The calves are born about the time poplar trees have leaves the size of squirrels' ears. There is one thing that somewhat puzzles me," continued Brown, "and that is, what becomes of the horns of the moose, which are shed every year. Big bull moose have horns like the tops of trees, yet I have found but few horns that had been shed, and I have been all through the woods during January and February, which is the time they are shed."

"Moose cover horns up," said Maypuck. "Moose paw moss, dirt, sticks, and snow on horns."

"Those same horns are mighty queer," said Brown. "It seems to me that no two sets are alike."

"Well, you are about right there," replied the naturalist. "Moose horns **are** very irregular. I have eighteen sets and they would furnish material for a small book on the subject."

"There is one thing about the moose that is always the same," said Brown, "and that is that there is no part of him that is lost when an Indian gets hold of him."

"Moose nickoshin," replied Kakagens. "Meat good, nose good, skin good, bone good, all heap good."

Dyche had already discovered that this assertion was correct, for none of the animals which had been killed had been wasted. The head is prized as an especial delicacy. The muffle is tender and even more sweet and delicate than a beaver's tail. Jerked moose-meat is much prized by the Indian and is similar to jerked buffalo-meat, but is not so tough. The hides make strong buckskins, from which the Indians make moccasins. A large hide will make from twelve to fifteen pair of moccasins, which are sold at a dollar and a half a pair. Hides of fœtal moose are much sought after by the Indians, who use them for making tobacco-pouches or tea-bags.

Moose-stories appeared to be the only subject that any of the party could do justice to, and every night the circle around the fire in the cabin was regaled with something fresh about the animal. In reply to a remark about "yarding," which is frequently described by Eastern writers, the Indians said that they had never seen more than seven or eight together in a band. As to the stories of hundreds of them assembled in a "yard," which is said to happen in Maine, the Indians shook their heads most energetically and said they had never seen it.

The season was now far advanced, December being half gone. The time for big snows had come, and

Dyche concluded that it would be well to get back to civilisation, where he could be in communication with home. Selling his ponies to a trader, he loaded his camp outfit and specimens on the wagon of a Dane who had come in after a load of fish and returned to Warren. There he found a number of letters from home, all telling of the illness of his boy. This news caused him to cut short his visit and hurry home, which he reached just six months after his departure in June.

CHAPTER XVIII.

In Colorado—On the Trail of Elk—A Night in the Snow—Deer, but no Elk—Another Wild-Goose Chase—The World's Fair King—The Last Hunt.

SEVEN months had passed since the return of the naturalist from the swamps of Manitoba—seven months of hard and wearing labor in workshop, laboratory, and class-room. The wind blowing through the trees of the university campus whispered enticing invitations to Dyche to lay aside again, for a brief period, his routine tasks and renew his energies by a sojourn in the mountains. With the invitation came thoughts of those who were his companions in the Cascades, and later he received two letters from these friends urging him to meet them in Denver and go with them to the mountains.

"Join us if possible," wrote the judge. "We want to be in the expedition that secures the big bull elk. We were with you when you secured the magnificent collection in British Columbia, and we think we can take you where you can get a World's Fair king. We can get Jim [Jim Kennicott, of Delta, Col.]. We had him last year and know him to be the right man. We have engaged him and six horses for ourselves, and the party will be complete with you. Shall we engage horses for you?"

A mountain king.

It only needed the stimulus of such a letter as this to cause the naturalist to drop all hesitation, and a month later, September 22d, Dyche, with the judge and doctor, started from Denver and met Jim, with his band of twelve horses, at Glenwood Springs. No time was lost here, and the train was moving up the trail that same afternoon.

For three hours the party climbed the mountain, but saw no sign of the cow-trail which had been indicated as a place to turn aside for wood and water and a good camping-place for the night. After nine o'clock a pond was found, and here the train was stopped and camp pitched. While the judge and Jim arranged the tent the doctor and Dyche went foraging for fuel, and soon returned to camp dragging what they supposed to be the top of a dead tree. When the supposed wood was thrown on the fire it was discovered that the dried bones of a horse would not burn, and a hearty laugh went round at the expense of the professor of anatomy and the physician who could not tell bones from wood. Nor did the joke fail to keep throughout the expedition.

Supper was followed by those countless tales of hunting adventure which are best told by the blazing camp-fire beneath the mountain pines. All next day they travelled through a beautiful mountain country until five in the afternoon, when a little park, in the middle of which stood two tall pines, was reached. Here was an ideal spot for a camp. That ubiquitous tin can, evidence of the march of civilisation, was found, and near the centre of the park was a miniature monument composed of the cans, bottles, and other relics of a party of campers. Rain and snow were falling when the tent was pitched, but such vagaries of the climate are unnoticed by true woodsmen, and after banking and draining the tent the safely ensconced party were soon telling their stories and cracking their harmless jokes at each other's expense. The doctor and judge had tried a cast in the **little** stream, but trout were shy, and they were unsuccessful in their efforts. Two **fine** bull elk had been seen crossing the park, **and this** was sufficient indication that the hunters were now in the elk country.

Marvine Lake, twelve miles farther on, was reached next day in a blinding snow-storm. Here **one** little trout was caught after the doctor and the judge had whipped the stream until their arms **ached.** There appeared to be many fish in the water, but they were not anxious to try the flies which were cast over them. Another day through rain and snow carried the party to a spot where the snow was cleared off sufficiently to permit the tent to be raised. While the judge and Jim were getting a fire and supper the naturalist and the doctor went hunting. They soon returned empty-handed, but with great stories of the deer they had seen and shot at.

At nine o'clock next morning the sun was shining and the pack-train was wending its way down Marvine River towards White River, which was reached shortly after noon. Here the horses varied the entertainment by giving an exhibition of what they could do in the way of bucking off the packs. One started it, and the contagion spread until five of them were rolling, bucking, and kicking at one time and the packs were in danger of complete demoralisation.

"Catch old Blackey! catch old Blackey!" excitedly called the judge as he ran breathlessly around the pitching and kicking squad.

"What's the matter with Blackey?" was the question as the other members of the quartette ran to his assistance.

"Why, the raspberry jam and pickles are in his pack."

The fears of the judge were only too well founded, for the jam and pickles became badly mixed, and the only consolation that was afforded the judge in his misfortune was the remark from the doctor that "they are all the better for that." This, however, did not meet the unqualified approval of the judge, who at once laid down a little law as to the conduct of the expedition. It took a full hour to get matters straightened out and all the rest of the day for the members of the party fully to express their opinions of the episode.

About the middle of the afternoon of the next day the party reached a spot in the heart of the mountains near Pagoda Peak which seemed suitable for a prolonged stay. After dinner Jim undertook to arrange the camp while the others went out to look for elk. The judge went east and the doctor north, while Dyche went through the heavy forest and brush to the northwest. After about two miles of slow travel, he found the fresh track of an elk and began stalking carefully. While moving silently through the underbrush he heard the animal start and run. Peering cautiously from his place of concealment, he saw it going at full speed up a ridge about four hundred yards away. Wondering what had startled the elk, the naturalist went to the spot where it had been standing and found the doctor's tracks. The animal had evidently seen or heard the hunter and had not waited to investigate. Darkness now came on and Dyche returned to camp, where he found the doctor and Jim, but the judge had not been heard from. Supper passed and still no judge, and then signal-shots were fired. For an hour this was continued, and then the answer was heard and soon the missing man put in an appearance, almost dead from fatigue. He could barely walk, but was wild with bull-elk fever.

"See that blood?" he asked, holding up a snow-ball on which were stains of red. "That came from a bull elk which I shot."

"Did you kill him?"

"I don't know. I tried to. I saw two bulls and tried to down one of them. I shot twice and then followed the bloody trail until dark. If I had not heard you shoot I would not have seen this camp to-night."

After a hasty breakfast next morning the whole party started out to help the judge find his elk. Dyche soon separated from the others and followed the fresh track of a mountain lion. The animal was going straight north towards Pagoda Peak. At one place in the snow was a spot where the lion had made a short run followed by several short jumps and then a long spring of at least thirty feet.

He had been after a deer, but the tracks in the snow showed that he had missed his dinner.

The naturalist now climbed to the top of Pagoda Peak, a great dome-shaped mountain with steps, or benches, on all sides, and took a survey of the country. All around him were mountains with valleys between. A band of about a dozen elk had been there before him, leaving their fresh tracks in the snow. According to Jim, it is the habit of elk to climb to the top of the highest peak in the neighbourhood and survey the whole country, locating the fires of hunters and then laying their plans for the day in such manner that they would keep away from their enemies.

The clouds, which had **been** growing heavier all the morning, were now sending down snow. Dyche followed along the ridge for two or three miles, and **during a lull in** the storm saw a band of elk feeding in a grassy park down the mountain-side. The wind was blowing directly towards the animals, **but** the hunter determined **to** get close to the band, **if** possible, and made a long stalk **around to** the other side of the park. Reaching a spot within a **hundred yards of** them, he lay hidden and began an examination with his field-glass. There were eleven **in all,** one young bull and the others cows and calves. The bull had a perfect set of horns, but as the beams were light and the animal of only ordinary size, the naturalist determined to let this band go and trust to the future to get the exact animal that was wanted. For two hours he lay in the snow, making notes of the movements and postures of the animals. He noted the carriage of the head and neck and ears and got a good idea of the appearance of the elk at home. One old cow grazed to within fifty yards of Dyche and then lay down and chewed her cud. Then the bull walked over that way and stood even closer. The temptation to shoot was very great, but the naturalist resisted it. Suddenly three other elk trotted into the park, seemingly under some excitement, which was communicated to the band and the whole lot went out of the opening at a round trot, moving off up the mountain and making a beautiful picture.

A depressed feeling came over Dyche as he saw the band going away, and he felt as though something, he knew not what, had gone wrong. Returning to camp he found the doctor there, but the judge and Jim were nowhere in sight or within earshot. The doctor was sure the men were lost, as they were still going north through the heavy snow-storm when he had left them early in the afternoon.

"We gave up that elk after travelling through heavy forests and much down timber, and I came back to Pagoda Peak and got my bearings while they kept going. They will have trouble before they see camp again. But it's all the better for that. Jim knows the mountains and the judge will learn after a few more experiences."

The doctor's prediction was verified, for several hours passed with no sign of either of the missing men. Signal-shots were fired at intervals and finally an answer was heard, and then the judge came stumbling into camp about ten o'clock. He was greatly exhausted, and it took the doctor and Dyche half an hour to bring him around to his normal condition. A pint of coffee and a warm supper, however, helped materially. The judge had been thoroughly bewildered in the storm and lost all bearings. He had become separated from Jim and did not know where the latter was. The continued absence of the hunter caused much uneasiness, and signal-shots were fired until long after midnight with no response from the guide. Morning dawned upon an uneasy trio. Jim had not returned, and it was now feared that he might be lying somewhere out on the bleak mountain badly hurt.

The doctor and Dyche were just preparing to make a search of the mountains when Jim arrived, too tired to talk. He presented a most woe-begone appearance and was completely worn out with the exertions of the previous night. He was so far gone that his face had assumed a deep copper colour and his hands trembled from the strain. A cup of hot coffee, dry stockings and shoes, and careful attention revived him somewhat, and then the doctor said it would be safe to feed him. Jim had been fasting for twenty-four hours, and he soon showed that his exhaustion did not extend to his appetite. Hot coffee, warm biscuit, and venison soon refreshed him and he was ready to tell the story of his night's adventures.

"Nobody knows where I stayed last night," he said. "I must have travelled over a hundred miles. I kept going as long as I could hold out and then made a fire and sat up with my back to a tree until morning. If it had not been for the professor's heavy Norwegian jacket I should have frozen. I'd have given ten dollars for a cup of coffee this morning."

"I think you'd better adopt my plan," said the professor, "if you are going to get caught away from camp much more. I always take a little coffee or tea in my haversack, with a cup and a few biscuit. If you would do that you could always make yourself at least half-way comfortable when you are caught out."

"I've got a better plan than that," was the reply. "You can punish me if you ever see me out of sight of this camp again while we are on this expedition, unless I'm after the horses."

"Do you know what the professor and I have decided to do with you and the judge?" asked the doctor.

"It would be hard to tell what you'd do."

"We've decided to put a bell on you and hobble the judge, and then we'll not have so much trouble worrying over you tramps."

"Well, I believe it would be a good scheme, and if you ever catch me out of sight of camp again you can bell me," said Jim.

The judge had been silently dressing his bruised feet during this conversation, and now held up his ankle, which showed marks similar to those on a hobbled horse, and remarked:

"See there! that foot looks as if it had been wearing hobbles for a week, and I'm sure it will hobble for a week more."

As it had taken forty-six shots to bring in the men, the camp was called Camp Forty-six, and then three days were spent in rounding up the horses and waiting for the judge and Jim to become able to travel. So much shooting had scared the game from that immediate vicinity, and it was decided to find new hunting-grounds.

A day's journey over mountains difficult of ascent carried the party across the range. Just before they reached a suitable camping-place a young buck jumped up and started to run, but a bullet from Dyche's rifle cut short its career. A day was now spent in examining the country for signs of elk. A trip of eight or ten miles west, through a section interspersed with numerous little grassy parks, showed that few hunters ever reached this part of the mountains. Deer stood in herds, almost as gentle as domestic cattle. More than a hundred were seen in half a day, and the presence of man did not appear to disturb them in the least. As Dyche was not after deer he did not kill any, but devoted much time to watching their actions and pose as they stood or fed in the parks.

As there were few signs of elk it was determined to retrace the trail over which they came and return to the country north of Pagoda Peak. The horses got through the forest with great difficulty owing to the snow, which caused them to slip and stumble. The progress was slow, but a beautiful little park was finally reached. It was hidden away in the mountains and was an ideal spot for a camp. The tent was pitched and preparations were made for a long stay. All were up early next morning, and as soon as breakfast was finished all except Jim went on a hunt for elk. The doctor and Dyche returned, unsuccessful, about noon, but the judge did not come in. His feet were still in poor shape for travelling, and he had intended to make but a short hunt. Jim now began to worry over his absence, and as the afternoon wore on with no return of the wanderer, the uneasiness was shared by the naturalist and the doctor. The latter walked along the base of Pagoda Peak in the direction taken by the judge, while Dyche went down a small stream along the valley. A little curling smoke about a mile below attracted the attention of the naturalist, and, in the hope that it might be the judge camped for the night, he made his way to it, reaching the place after dark. There was a deserted camp-fire smouldering at the end of a log, while a few spruce boughs piled against a tree indicated that someone had spent a night there not long before. After an examination Dyche became convinced that this was the spot where Jim had spent the night when he had walked that "hundred miles." His surmise was proven correct when Jim afterwards saw the place and recognised it.

Not a sign of the judge could be found, and it was with a feeling of great anxiety that preparations were made for the night. Signal-shots were fired at intervals until midnight, but there was no response. Morning dawned upon a trio of very anxious hunters. A council of war was held, and it was decided that a general search should be made if the missing man did not return by eleven o'clock. Dyche could not rest until that time, and so **he** took his gun and went for a short hunt. As he returned he recognised the judge's familiar form seated by the fire. The judge said he had been out in the woods listening to a bull elk bugling.

"I went to the top of the peak, and after taking a good look at the country started back to camp. I got on the wrong side of the ridge and went **about** three miles northeast, when I should have gone northwest. About the time I discovered that I was going wrong I heard a bull elk blowing his bugle down in the valley, and this made me forget everything else. I followed the noise but failed to obtain a glimpse of the animal, and by that time it was growing dark. I was in a deep spruce forest, and you know how easy it is to get mixed up in such a place. My feet were in such a condition that I did not propose to go wandering around in the dark. I was not lost, but just too tired to go to camp. I made me a bed and kept the owls company for the rest of the night. That elk was bugling all night long. When daylight came I went **back** to the top of the peak and soon located camp."

"That's a good story," said the doctor, "but he don't tell all of it. Jim and I started out for Pagoda Peak this morning, thinking we might find the judge. We had gone but a little way when we saw a man crawling along. We at once thought that the judge had been badly hurt and had been crawling back to camp all night. We soon found, however, that the man was making a stalk on one of the horses, and if we had not rushed up just in time he would have shot one, **sure**. We disarmed him and brought him to camp, where venison steak, hot coffee and biscuit soon brought back his reason. But he's all the better for that. There's many a worse man than the judge—in the penitentiary."

"Begorra, I can't say as much for the doctor," was the judge's sole reply.

Snow fell all day Sunday and the day was spent in camp. Monday found a fine "tracking" snow on the ground. Dyche was anxious for a hunt, but the other members of the party were anxious to get away from "Lost Park," as the place had been named. The altitude was too high and camp was too hard to find to suit them. As Dyche had brought them into the place it fell to him to get them out. The train was packed and started, the naturalist in the lead. A lion's track in the snow was too great a temptation to resist, and the leader of the cavalcade followed off after the beast. A short distance farther on an elk winded him and started through the woods. The tracks indicated that it was a big bull, and as that was just what he had come for, he took up the new trail. Feeling that the other members of the party would understand his purpose when they saw his tracks crossing those of the elk, he started to hunt it Indian fashion, as he had been taught in the swamps at the Lake of the Woods. Taking up a line about two hundred yards to the leeward, Dyche slipped along through the woods towards a point where he was confident the animal would cross to the other side of the cañon. Going carefully to the bottom of the cañon, he went towards the trail and found where the bull had crossed over. On a little hill were signs that the bull had stopped and looked back over his trail and then, seeing that he was not followed, had moved off more leisurely. A long grove of quaking asps led up the side of the mountain along a branch of the Williams River, and up this branch Dyche crept, carefully, towards a grove. Just as he raised his head to take a look between the trees the elk started at a round **trot**. Quickly throwing the Winchester forwards, the naturalist sent a ball after the animal, which had the effect of accelerating its pace. A second shot was fired which caused the bull to break into a run. Now it came into plain view and a third shot was fired. With the report the animal disappeared over the ridge.

Hastening to the spot where the elk was last seen, Dyche found the magnificent creature lying dead. He was a grand specimen and well fitted to lead the group which had been begun for the World's Fair. His standing height was just sixty-three inches, while his girth was seventy-five inches just behind the shoulder. His girth about the abdomen measured eighty-three inches. He was not phenomenal in size or length of horns, but his proportions were perfect and his skin was at its very best, being especially rich in colour. By the time the other members of the party reached the spot all the measurements had been taken and the naturalist was skinning his prize. As the pack-train came in sight the doctor gave a cheer. His first question showed what was uppermost in everyone's mind.

"Is he a World's Fair king?"

An affirmative answer brought out a chorus of hurrahs and a storm of congratulations. The pack-train was stopped and everyone assisted in skinning the elk. There was a little blood in the animal's mouth, but not a bullet-hole could be found in the skin anywhere. Thereupon Dyche was subjected to much chaffing for scaring an elk to death. These quizzing remarks continued until it was found that a bullet had passed directly through the heart. The glory of the exploit was enhanced a short time after, when the naturalist shot a badger without drawing blood; the bullet struck the rock directly under the animal, killing it by the concussion.

Camp was made two miles farther east on the north branch of Williams River. A high mountain rose to the north. To the west was the range over which they had just come and which was made beautiful by the vast forests of spruce and fir. To the east and south the country opened out into a valley, giving a view of ten or fifteen miles of changing lights and shades through the openings in the mountains. The judge at once christened the place Camp Grand View.

Next day, while Dyche was preparing the skeleton of his specimen, the doctor, who had accompanied him, rambled off on a little hunt. The naturalist saw a band of elk emerge from the timber and walk towards the doctor, who hid in a ravine at the foot of a little ridge. A number of cows and calves were at the head of the band, and as they came down the ridge they smelled powder and turned down the side of the mountain into the woods. Following the cows and calves came two fine bulls. They walked on down the ridge, not suspecting danger, in plain sight of Dyche. The latter was just wondering what had become of the doctor, when he saw a puff of smoke and heard the gun crack. One of the elks began running in a circle and soon fell to the ground, while the other disappeared in the timber. It was a finely proportioned animal with a perfect set of horns and a good growth of hair.

Snow had been falling now for twenty-four hours, and it was possible to hunt without becoming confused by old trails. Dyche started out early next morning to see what was in the country. He found a bear-track and followed it, even though he knew that he would have little chance of getting it. The bear visited no less than four carcasses of elk lying in the snow, all cows and calves. These had been killed by hunters and left lying just where they fell, not a pound of flesh being taken or an inch of skin cut off. The bear-hunters go through the country killing deer and elk, leaving the carcasses for bait. **If** a bear visits one of them a week or a month later, the hunters set their traps.

The camp of some bear-hunters was found, and Dyche went with one of them to look at the remains of a bull elk. The specimen was spoiled and useless to the naturalist except the leg bones and skull. During that day Dyche saw eleven bodies of elk which had been slain for bear-bait. These two bear-hunters were novices and appeared to be out for the fun of the trip more than anything else. They gave the naturalist a cordial invitation to spend the night with them. As he was a long way from his own camp and night was coming on, it took but little pressing to induce him to accept.

The men gave their names as Henry Maybe and Henry Geisler and they had travelled many miles in following bear. They were ready to testify that a bear can travel twenty or thirty miles a day without stopping to rest or eat. After the usual stories of hunting and adventure in the mountains the camp became silent. Dyche had barely fallen asleep when he was aroused by the voice of one of his companions.

"Henry, what is that?" came the startled whisper from one roll of blankets.

"That sounds just like that bear cub we saw at Meeker," was the reply.

"Let's get up. Maybe we can catch it."

Hastily scrambling from their blankets, the two went into the woods in the direction of the noise. Dyche had lost no bear cub and remained in his blankets. The men returned, wondering what it could have been. They had barely become settled in their blankets when the sound came again. They hastened to the woods with the same result. They now determined to exercise a little ingenuity and surprise the animal. One man remained in the woods while the other returned to the tent. Soon the sound was heard again and the watcher called out:

"Hurry up, Henry. Here it is. It's up in a tree."

Henry took his gun and a lantern and for a while all was quiet. After an absence of several minutes they returned to camp, using strange words about "that consarned little owl."

Returning to Camp Grand View next day, Dyche found that everyone had reached the conclusion that it would be well to leave the place, and several days were spent in travelling and searching for a more suitable spot. Darkness compelled them to stop one night on a high point on the mountains, where the wind blew a perfect gale. The judge gave it the name of Camp Windy Point. A few days were spent here, and Jim made a short expedition and reported that he had fired at a large bull elk but had not drawn blood. Next day the doctor and the judge went out on horseback while Jim and Dyche went together on foot. Jim showed where he had seen the elk, but there were no signs of a wounded animal. The trail was followed some distance and then it was decided to return to camp.

After travelling awhile Dyche sat down for a short rest at the edge of a grassy park. The bugle-call of an elk off in the woods aroused him, and soon he heard a response in an opposite direction. This bugle-call of the elk is perhaps the most peculiar sound emitted by an animal. The elk starts off with a noise similar to that of the squeaking of an old barn door, and this harsh noise is followed by a bawl as of a cow. This develops into the bray of a mule and winds up with a peculiar, clear bugle note or call, which rings full and free through the woods. It is a sound which never fails to excite hunters and set their blood to bounding. The naturalist and Jim were just preparing to begin a stalk when an elk emerged from the woods on the opposite side of the park. It was an old cow, and soon others appeared. They trooped out of the woods until fully sixty were in sight. Then from another point, off to one side, thirty more came out. All these were cows and calves, with a very few young bulls. Now the old bulls began to make their appearance, the last to come into the park. This immense band was counted, and it was found that there were one hundred and twenty-five.

The wind was blowing directly from the hunters towards the band, and a consultation showed a diversity of opinion as to the proper mode of procedure. Jim was a good woodsman, but the sight of so many elk at one time evidently "rattled" him and gave him an attack of elk fever. He strongly opposed Dyche's plans for a long stalk which would take them around and ahead of the band where they could lie in ambush and select the finest of the lot, but insisted on crawling under poor cover directly down towards them. He was afraid to let them get out of his sight lest they should go off and never be seen again. Much against his will, Dyche consented to try Jim's plan, for he saw that any other course might cause the excited hunter to make an untimely movement which would spoil the entire stalk.

Jim's plan was followed, and then it was found that they had to retrace their steps and begin over again. Three times was this done before Jim became convinced that Dyche's idea was the better. But now it was almost too late, for the band had fed near the spot where the naturalist desired to hide. The animals were now between two ridges, moving slowly. If the hunters could get ahead of the band they could examine the elk at their leisure and see if the one they wanted was in the herd. If there was a bull larger than the one killed by Dyche, he wanted it to lead the World's Fair group. If not, the naturalist preferred that none be killed, for he had a sufficient number of specimens of all other kinds.

After losing much time in following out Jim's vagaries they reached the spot where Dyche desired to go at the beginning of the stalk, but they had been so long on the way that the animals were there ahead of them and were passing into the ravine. As the hunters crawled through the underbrush, their ears were assailed by a queer combination of grunting and squealing, made by the elk as they fed along through the park. Now and then a bull raised his head and gave his bugle-call, and the response came from the far side of the band. The hunters at last reached a point about a hundred yards from the game, but the trees and brush were so intertwined that it was impossible to see beyond them. Occasionally a head or a portion of the body of an elk came into view, but it was evident that the men must get closer if they wished to see the entire band. Slipping off their shoes and jackets they crawled slowly forwards for about fifteen yards. Suddenly Jim jumped to the top of a log which barred the way and was in full view of the band of elk.

About fifty cows and calves were feeding directly towards him, and as he appeared on the log the animals started on the run, frightening the others as they went through the park at full speed. Jim began shooting at the fleeing elk as fast as he could work the lever of his Marlin. Dyche could see nothing, and in trying to get around the log where the way would be clear, he lost considerable time. Jim went on the run through the woods and Dyche followed, reaching the edge of the timber just in time to see the band disappearing at the lower end of the ravine. Jim was running at full speed down the ridge to the right of the ravine and Dyche went on the one to the left. After the first alarm the elk bunched and were running together in a compact mass. Into this bunch Jim now began shooting at a distance of only fifty yards. He worked his gun as rapidly as possible and poured balls into the band like leaden rain.

Dyche carefully surveyed the band and singled out a bull which appeared to be much larger than any of the others. Towards this animal he turned his attention, shooting at it until it left the band and went into the woods alone. The band had now gone around the hill on which Jim was standing, and Dyche heard more shooting and then all was still.

"Prof, why ain't you shootin'?" shouted the excited mountaineer.

The naturalist was disgusted and mortified and yet amused at the way the hunt had ended. Only one elk was needed and that one must be a monster, but Jim had been indiscriminately slaughtering the animals at short range, when there was no possible way of utilising the flesh.

Dyche now hurried across to the place where Jim

had been standing and saw the latter about four hundred yards farther on and the elk in a bunch about seventy-five yards below him. Jim had used his last cartridge and was watching the band, which ran about a hundred yards farther and then stopped on a knoll. Another run took them to the woods and then they were out of sight. Jim went back after the shoes and jackets and told the naturalist of the number of big ones he had killed. It was decided to go back to camp, as it was almost dark, and return in the morning for the dead elk.

The evening was spent by Jim and Dyche in telling over and over again what they had done and what a lot of fine bulls would be found dead. Early in the morning the whole party went to the field of action, expecting to find at least a dozen dead elk. A very careful search revealed the fact that not an elk had been killed, while the only evidence that any had been wounded was an occasional drop of blood, which might have come from a sore foot or an injured leg. The trail of the fleeing elk was followed four or five miles, and then Jim was compelled to acknowledge that he had missed the whole herd. On the return the judge rounded up a fine bull which had straggled from the band, evidently the one at which Dyche had shot, and took it into camp, thus ending their great elk-hunt in the Rockies.

Every member of the party was now satisfied with the result of the trip to the mountains, including Jim, who, however, was a little sore over his experience with the big band, and the party was ready to return to civilisation.

"How large do elk grow?" asked the doctor as they were packing the specimens and commenting on the big fellow's size.

"A full-grown bull sometimes attains the weight of a thousand pounds, but the average is not over half of that. Cows weigh from three hundred to five hundred pounds and have but one calf, which is spotted like a fawn."

"If the skin was as good as the meat," said Jim as he helped himself to another rib, "it would make mighty good leather, but it's the poorest part of the animal."

"That's a fact," replied Dyche; "the skin is almost worthless and appears to be rotten when tanned."

"Where is the best place to find elk now?" asked the judge.

"The range is now restricted to certain favoured spots in the Rocky Mountains, but they formerly ranged from Northern Mexico to British Columbia. A few are found in Minnesota. They live principally on grass and twigs and sometimes bits of the bark of such trees as quaking asps. The European cousin of the elk is undoubtedly the red deer, or stag. Elk will average much larger than the stag, but they are similar in general appearance and when placed in parks together will breed and their offspring will also breed."

"What a fine team a pair of them would make!" said Jim. "They would take a fellow sailing over the country in great style."

"I've seen that done," said the judge. "Last year it was a common sight in Denver. A man had a pair of five-year-old bulls hitched to a cart and drove through the streets."

"Yes," said Dyche, "they are easily broken to harness and seem to take kindly to it, but when they take a notion to run away there is sure to be a general smash-up."

The first camp on the return to Denver was made at Trapper's Lake, and while hunting near the water Dyche saw hundreds of trout disporting themselves. He threw stones at them until his arm was tired, and then went to camp with his story and was laughed at for his pains. The judge looked at the professor quizzically, and the doctor said he had been pretty good at fish-stories himself in his day. Dyche took the chaffing and also the judge's fishing-rod and returned to the lake, the doctor following out of curiosity. The water was fairly alive with the speckled mountain trout, and the lines could hardly be thrown in fast enough to satisfy the fish. At the end of an hour of this sport they returned to camp with seventy fine fish in a sack. They weighed just fifty pounds, and the judge, who was admittedly the best fisherman in the State, said it was the finest catch he had ever seen.

The month had been most pleasantly spent by the whole party, and the camp-fires of a naturalist were things of the past. At Denver the naturalist parted from his friends, promising that at some future day they would meet and take another and longer hunt in the mountains.

CHAPTER XIX.

Results of the Camp-Fires—The Specimens Obtained.

THE camp-fires are dead and the ashes cold. Hundreds of whitened spots surrounded by burned and blackened wood and by bleaching bones may be found from the sand-hills of No-Man's Land and the bleak prairies of western Kansas to the marshes, bogs, and fens of Manitoba; from the pine-covered ridges at the head of the Pecos in New Mexico to the deep forests on Kettle River in British Columbia, where silence reigns supreme. Mountain and plain, swamp and lava-bed have been called upon to contribute their quota, and the work of the naturalist is now changed from active field operations to the preparation of specimens in the laboratory.

It has not been my purpose to make a book of thrilling adventures, full of exciting hunting-stories, but rather to set forth clearly and truthfully those incidents which impart a peculiar charm to the life of hunters who go to the woods with the true sportsman's instinct and look for game to add something to the knowledge of the world. That the naturalist had many more adventures than those recorded in these pages can readily be seen by a walk through the storerooms of the Kansas University.

Here, piled in apparent confusion, but really classified with the greatest care and labelled so definitely that no chance of error is left, are to be found the thousands of specimens which have been obtained not only by his own rifle, but also by exchange and by purchase from those who spend their lives in the wildwood haunts of the fauna of the continent. Trappers, hunters, traders, prospectors, and Indians have been drawn upon for specimens, as well as for incidents showing the peculiarities of each species, and the result is that many specimens of rare, and in some instances extinct, animals have been preserved for the study of future generations.

In connection with this subject it should be remembered that a complete series must embrace a whole family in all stages of development, in order to show all the characteristics and peculiarities of the animal. Not alone are the old male and female specimens necessary, but the calves and young of different stages must be shown to give a good idea of the species. Often it has been necessary to collect specimens from widely separated localities in order to exhibit the peculiar differences brought about by environment.

While the collection now stored at the university may appear large and beyond the requirements of science, it must be remembered that in many instances these specimens will probably, in the near future, be the only representatives of those rare species which are rapidly becoming extinct. Eight years ago only four specimens of the Rocky Mountain goat were to be found in all the museums of the world.

These were in such condition that they told nothing either of the appearance or the peculiarities or habits of the animal as he is seen in his mountain home. So much hardship attends the hunting of this most rare of all the North American fauna that it is with the greatest difficulty that a single specimen can be obtained, and then the additional labour of getting the game down the almost inaccessible mountain which the animal has chosen for its habitat makes the securing of a single perfect example an event worthy of being chronicled in the scientific journals of the world. The thirty-four complete specimens of this animal secured by Professor Dyche will go far towards aiding the future naturalists in their studies.

While all the expeditions of the naturalist were for the purpose of collecting examples of the larger mammals of the continent, he never lost sight of the fact that he visited places where he was likely to find rare forms of smaller mammals and birds, and he secured thousands of them which, with his larger specimens, form a collection that for excellence cannot be equalled in the whole world.

A list must of necessity be incomplete, but a partial enumeration of the rarest forms will give the reader some idea of the wonderful work that has been accomplished during these expeditions. The American bison is now practically extinct as a wild animal. The few remaining herds are so hedged about by law that they are practically domesticated. Fourteen complete specimens of these were secured from the last survivors of the wild herd. Of antelope he has thirty-five specimens, of bear fourteen, mountain lions eleven, elk **nineteen, Rocky Mountain** sheep **forty-five,** Rocky Mountain goats **thirty-four,** moose forty, deer **sixty,** timber-wolves **ten,** coyotes twenty-four, foxes twenty-five, including some of the rarest silver **and** cross varieties, lynxes ten, caribou seven, and of otter, beaver, wolverine, fisher, badgers, woodchuck, raccoons, and smaller animals several hundreds.

While the money value **of** the collection has **never** entered into **the** consideration **of** the professor, **who,** with the true instincts of a naturalist, has been wholly **engrossed with** the collection **of** the specimens, **it would now be impossible** to duplicate the collection **for $100,000, and** as the animals become scarce and **rare** this value will increase until within a very few years it will be of incalculable worth. Realising that **future** generations may look **to this** collection **for** the study of rare species, the professor aims to have **in** addition to the mounted groups, which will stand as silent educators **to** the masses, a student's series which will be of **use to** the future naturalist who may have no other means **of** examining an extinct species.

Nor is **the work complete as the** collection **now stands.** The material on hand is to be increased **by the results of** other expeditions now in contemplation, **and** these specimens are to be the basis of works **on the** natural **history of North** America, the first of which, **on** ruminants, **is now in** course **of preparation.**

Other camp-fires will glow in the deepest wilds of British-Columbian mountains while the naturalist **will** seek rare **and** almost extinct species of that region. Far-off Alaska and **even** Greenland will be called upon to give up tributes of musk-ox and polar bear, while Mexico and the Gulf coast will be drawn upon for phases of animal life peculiar to those sections.

The ashes of the camp-fires from the Sierra Madre and Sangre del Christo to the Cascade ranges, from the Columbia River to the Lake of the Woods, now mark the spots where many happy hours were spent while communing with Nature in her most secret haunts. The sparks are extinct and many rains and snows have beaten the white ash into the cold earth, yet there remain the bright memories which he alone can know who goes to the deep solitudes and sleeps beneath the singing pines.

We have taken you through the mountain, prairie land, and swamp; we have shown you the discomforts as well as the pleasures of the life of a naturalist; we have given you a hunter's fare of venison and camp bread, washed down with black coffee made in a tin cup; we have laid you to rest on a bed of spruce boughs and sung you to sleep with the sighing of the wind as it plays through the tree-tops and rustles among the pine-needles. If we have given you one desire for that free life in the woods and mountains or added one iota to your pleasure, we are repaid.

THE END.

www.ingramcontent.com/pod-product-compliance
Lightning Source LLC
Chambersburg PA
CBHW031857220426
43663CB00006B/667